NUREG-1617

Standard Review Plan
for Transportation Packages for Spent Nuclear Fuel

I0484553

Final Report

Manuscript Completed: January 2000
Date Published: March 2000

Spent Fuel Project Office
Office of Nuclear Material Safety and Safeguards
U.S. Nuclear Regulatory Commission
Washington, D.C. 20555-0001

ABSTRACT

The Standard Review Plan (SRP) for Transportation Packages for Spent Nuclear Fuel provides guidance for the review and approval of applications for packages used to transport spent nuclear fuel under 10 CFR Part 71.

This document is intended for use by the U.S. Nuclear Regulatory Commission (NRC) staff. Its objectives are to (1) summarize 10 CFR Part 71 requirements for spent fuel transport package approval, (2) describe the procedures by which NRC staff determines that these requirements have been satisfied, and (3) document the practices used by the staff in reviews of package applications.

This NUREG is expected to be updated on a periodic basis. As issues arise between updates, the Spent Fuel Project Office will issue interim staff guidance (ISG) where the SRP guidance needs revising. ISG's are placed in the NRC public document room and on the NRC WEB for public information. ISG's, as issued, replace specific portions of the SRP. Comments regarding errors or omissions, as well as suggestions for improvement of this NUREG and subsequent ISG, should be sent to the Director, Spent Fuel Project Office, U.S. Nuclear Regulatory Commission, Washington, DC 20555-0001.

CONTENTS

FIGURES

TABLES

ACRONYMS AND ABBREVIATIONS

ALARA	as low as is reasonably achievable (radiation exposure)
ANS	American Nuclear Society
ANSI	American National Standards Institute
ASME	American Society of Mechanical Engineers
ASTM	American Society for Testing and Materials
B&PV	Boiler and Pressure Vessel (ASME Code)
Bq	Becquerel
BWR	boiling-water reactor
°C	degrees Celsius
Ci	Curie
CFR	U.S. Code of Federal Regulations
DOT	U.S. Department of Transportation
°F	degrees Fahrenheit
g	gravitational unit
k_{eff}	"k" effective-neutron multiplication factor
LSA	low specific activity
MIL	military
MNOP	maximum normal operating pressure
mrem	millirem
mSv	millisievert (1 mSv = 100 mrem)
NMSS	NRC Office or Nuclear Material Safety and Safeguards
NRC	U.S. Nuclear Regulatory Commission

PWR	pressurized-water reactor
RG	regulatory guide (NRC)
SAR	safety analysis report
SCO	surface contaminated object
SER	safety evaluation report
SI	International System of Units
SFPO	Spent Fuel Project Office (NRC NMSS)
SNF	spent nuclear fuel
SRP	standard review plan
SSCs	structures, systems, and components
Sv	Sievert

GLOSSARY

The following terms are defined here by the staff for the purpose of this SRP. Many of the terms are taken from 10 CFR 20.1004, 10 CFR 71.4, or 49 CFR 173.403. The definitions from these CFR sections have not been changed in the list below, but are repeated for convenience. Standards are expressed in the International System of Units (SI). The U. S. standard or customary unit equivalents presented in parentheses are for information only.

A_1	the maximum activity of special form radioactive material permitted in a Type A package.
A_2	the maximum activity of radioactive material, other than special form, LSA and SCO material, permitted in a Type A package.
As low as is reasonably achievable (ALARA)	making every reasonable effort to maintain exposures to radiation as far below the dose limits in this part as is practical consistent with the purpose for which the licensed activity is undertaken, taking into account the state of technology, the economics of improvements in relation to benefits to the public health and safety, and other societal and socioeconomic considerations, and in relation to utilization of nuclear energy and licensed materials in the public interest.
Becquerel (Bq)	a unit, in the SI, of measurement of radioactivity equal to one transformation per second.
Benchmarking	validation of the accuracy of a computer code by comparison of obtained results with those of previously determined experimental values.
Bias	For criticality calculations, ANSI/ANS-8.1 defines bias as a measure of systematic differences between calculations and experimental data and subsequently defines uncertainty in the bias. See NUREG/CR-6361 for further discussion of bias. The determination of bias must adequately consider the variation in the differences between the calculations and experimental data.
Carrier	a person engaged in the transportation of passengers or property by land or water as a common, contract, or private carrier, or by civil aircraft.
Certificate holder	a person who has been issued a certificate of compliance or other package approval by NRC.
Certificate of compliance	a certificate issued by NRC which authorizes the use of a specific packaging, for a specified time, and for a specified scope of activity.
Close reflection by water	immediate contact by water of sufficient thickness for maximum reflection of neutrons.

Closed transport vehicle	a transport vehicle or conveyance equipped with a securely attached exterior enclosure that during normal transportation restricts the access of unauthorized persons to the cargo space containing the Class 7 (radioactive) materials. The enclosure may be either temporary or permanent, and in the case of packaged materials may be of the "see-through" type, and must limit access from the top, sides, and bottom.
Code	used generically to refer to national or "consensus" codes, standards, and specifications, or specifically to refer to the ASME Boiler and Pressure Vessel Code or may be used to describe computer models.
Confirmatory calculations	calculations made by the reviewer to determine whether the package design and specifications meet the regulations. These calculations do not replace the design calculations and are intended to assess and confirm the basis and conclusions of the applicant's calculations.
Containment system	the assembly of components of the packaging intended to retain the radioactive material during transport.
Conveyance	for transport by public highway or rail, any transport vehicle or large freight container; for transport by water, any vessel, or any hold, compartment, or defined deck area of a vessel, including any transport vehicle on board the vessel; and for transport by aircraft, any aircraft.
Curie (Ci)	the basic unit to describe the intensity of radioactivity in a sample material. A curie is equal to 37 billion disintegrations per second.
Damaged spent nuclear fuel	spent nuclear fuel with known or suspected cladding defects greater than a hairline crack or pinhole leak.
Depleted uranium	uranium containing less uranium-235 than the naturally occurring distribution of uranium isotopes.
Docketed	formal submissions made to NRC by an applicant, and officially filed by NRC in the Agency's records for the application. NRC assigns a docket number to the transportation package, which is used for the application and subsequent submissions and other correspondence regarding the package. Except when NRC concurs in a request that material be protected as being "proprietary data," docketed material, in accordance with 10 CFR 2.790, becomes available for public copying.
Enriched uranium	uranium containing more uranium-235 than the naturally occurring distribution of uranium isotopes.
Exclusive use	the sole use by a single consignor of a conveyance for which all initial, intermediate, and final loading and unloading are carried out in

accordance with the direction of the consignor or consignee. The consignor and the carrier must ensure that any loading or unloading is performed by personnel having radiological training and resources appropriate for safe handling of the consignment. The consignor must issue specific instructions, in writing, for maintenance of exclusive use shipment controls, and include them with the shipping paper information provided to the carrier by the consignor.

Fissile material	plutonium-238, plutonium-239, plutonium-241, uranium-233, uranium-235, or any combination of these radionuclides. Unirradiated natural uranium and depleted uranium, and natural uranium or depleted uranium that has been irradiated in thermal reactors only, are not included in this definition. Certain exclusions from fissile material controls are provided in 10 CFR 71.53.
Fissile material package	a fissile material packaging together with its fissile material contents.
g	gravitational unit. (1 g = force exerted on a mass vertically by gravity)
Independent calculation	calculations separate from the applicant's. Input data should be taken from primary sources such as the package drawings and manufacturer's specifications. Models should be developed separately by the reviewer. To the extent possible, different techniques, codes, and cross section sets or other derived data sets should be used.
"k" effective	the ratio of the number of neutrons resulting from fission in one generation to the number of neutrons resulting from fission in the preceding generation.

Low specific activity (LSA) material	radioactive material with limited specific activity that satisfies the descriptions and limits set forth below. Shielding materials surrounding the LSA material may not be considered in determining the estimated average specific activity of the package contents. LSA material must be in one of three groups:
	(1) LSA-I.
	    (i) Ores containing only naturally occurring radionuclides (e.g., uranium, thorium) and uranium or thorium concentrates of such ores; or
	    (ii) Solid unirradiated natural uranium or depleted uranium or natural thorium or their solid or liquid compounds or mixtures; or
	    (iii) Radioactive material, other than fissile material, for which the A_2 value is unlimited; or
	    (iv) Mill tailings, contaminated earth, concrete, rubble, other debris, and activated material in which the radioactive material is essentially uniformly distributed, and the average specific activity does not exceed 10^{-6} A_2/g.

(2) LSA-II.

(i) Water with tritium concentration up to 0.8 TBq/liter (20.0 Ci/liter); or

(ii) Material in which the radioactive material is essentially uniformly distributed, and the average specific activity does not exceed 10^{-4} A_2/g for solids and gases, and 10^{-5} A_2/g for liquids.

(3) LSA-III. Solids (e.g., consolidated wastes, activated materials) in which:

(i) The radioactive material is essentially uniformly distributed throughout a solid or a collection of solid objects, or is essentially uniformly distributed in a solid compact binding agent (such as concrete, bitumen, ceramic, etc.);

(ii) The radioactive material is relatively insoluble, or it is intrinsically contained in a relatively insoluble material, so that, even under loss of packaging, the loss of radioactive material per package by leaching, when placed in water for 7 days, would not exceed 0.1 A_2; and

(iii) The average specific activity of the solid does not exceed 2×10^{-3} A_2/g.

Low toxicity alpha emitters	natural uranium, depleted uranium, natural thorium; uranium-235, uranium-238, thorium-232, thorium-228 or thorium-230 when contained in ores or physical or chemical concentrates or tailings; or alpha emitters with a half-life of less than 10 days.
Maximum normal operating pressure (MNOP)	the maximum gauge pressure that would develop in the containment system in a period of 1 year under the heat condition specified in 10 CFR 71.71(c)(1), in the absence of venting, external cooling by an ancillary system, or operational controls during transport.
Natural thorium	thorium with the naturally occurring distribution of thorium isotopes (essentially 100 weight percent thorium-232).
Natural uranium	uranium with the naturally occurring distribution of uranium isotopes (approximately 0.711 weight percent uranium-235, and the remainder by weight essentially uranium-238).
Normal form radioactive material	radioactive material that has not been demonstrated to qualify as "special form radioactive material."
Optimum interspersed hydrogenous moderation	the presence of hydrogenous material between packages to such an extent that the maximum nuclear reactivity results.
Package	the packaging together with its radioactive contents as presented for transport.

Packaging	the assembly of components necessary to ensure compliance with the packaging requirements of 10 CFR Part 71. It may consist of one or more receptacles, absorbent materials, spacing structures, thermal insulation, radiation shielding, and devices for cooling or absorbing mechanical shocks. The vehicle, tie-down system, and auxiliary equipment may be designated as part of the packaging.
Radiation level	the radiation dose-equivalent rate expressed in millisievert(s) per hour or mSv/h (millirem(s) per hour or mrem/h). Neutron flux densities may be converted into radiation levels according to Table 1, 49 CFR 173.403.
Radioactive contents	a Class 7 (radioactive) material, together with any contaminated liquids or gases within the package.
Radioactive material	any material having a specific activity greater than 70 Bq per gram (0.002 microcurie per gram).
Rem	the special unit of any of the quantities expressed as dose equivalent. The dose equivalent in rems is equal to the absorbed dose in rads multiplied by the quality factor (1 rem = 0.01 sievert).
Rule	unless used generically, a requirement stated in the Code of Federal Regulations.
Sievert (Sv)	the SI unit of any of the quantities expressed as dose equivalent. The dose equivalent in sieverts is equal to the absorbed dose in grays multiplied by the quality factor (1 Sv = 100 rems).
Safety analysis report (SAR)	in the context of this SRP, the report submitted by the applicant in compliance with 10 CFR Part 71, Subpart D. The fundamental contents of the report are described in 10 CFR 71.31. Guidance on format and content of the report is provided by Regulatory Guide 7.9, "Standard Format and Content of Part 71 Applications for Approval of Packaging for Radioactive Material." The SAR is considered to be the submitted application, along with any supplemental data and responses submitted to NRC staff to resolve questions arising during the staff's review. Only docketed material is considered to form part of the submission. The effective SAR is that submitted, as amplified and/or modified by the supplemental and later submissions.
Safety evaluation report (SER)	in the context of this SRP, the report prepared by NRC staff to document the acceptability of the applicants SAR and other required submissions. The SER also identifies NRC staff's conclusions and the conditions of approval that will be included in the certificate of compliance.

Special form radioactive material	radioactive material that satisfies the following conditions: (1) It is either a single solid piece or is contained in a sealed capsule that can be opened only by destroying the capsule; (2) The piece or capsule has at least one dimension not less than 5 mm (0.2 in); and (3) It satisfies the requirements of 10 CFR 71.75. A special form encapsulation designed in accordance with the requirements of 10 CFR 71.4 in effect on June 30, 1983 (see 10 CFR Part 71, revised as of January 1, 1983), and constructed before July 1, 1985, and a special form encapsulation designed in accordance with the requirements of 10 CFR 71.4 in effect on March 31, 1996, (see 10 CFR Part 71, revised as of January 1, 1983), and constructed before April 1, 1998, may continue to be used. Any other special form encapsulation must meet the specifications of this definition.
Specific activity of a radionuclide	the radioactivity of the radionuclide per unit mass of that nuclide. The specific activity of a material in which the radionuclide is essentially uniformly distributed is the radioactivity per unit mass of the material.
Spent nuclear fuel (SNF)	fuel that has been withdrawn from a nuclear reactor following irradiation, the constituent elements of which have not been separated by reprocessing.
Surface contaminated object (SCO)	a solid object that is not itself classed as radioactive material, but which has radioactive material distributed on any of its surfaces. SCO must be in one of two groups with surface activity not exceeding the following limits: (1) SCO-I: A solid object on which: (i) The non-fixed contamination on the accessible surface averaged over 300 cm^2 (or the area of the surface if less than 300 cm^2) does not exceed 4 Bq/cm^2 (10^{-4} microcurie/cm^2) for beta and gamma and low toxicity alpha emitters, or 0.4 Bq/cm^2 (10^{-5} microcurie/cm^2) for all other alpha emitters; (ii) The fixed contamination on the accessible surface averaged over 300 cm^2 (or the area of the surface if less than 300 cm^2) does not exceed 4×10^4 Bq/cm^2 (1.0 microcurie/cm^2) for beta and gamma and low toxicity alpha emitters, or 4×10^3 Bq/cm^2 (0.1 microcurie/cm^2) for all other alpha emitters; and (iii) The non-fixed contamination plus the fixed contamination on the inaccessible surface averaged over 300 cm^2 (or the area of the surface if less than 300 cm^2) does not exceed 4×10^4 Bq/cm^2 (1 microcurie/cm^2) for beta and gamma and low toxicity alpha emitters, or 4×10^3 Bq/cm^2 (0.1 microcurie/cm^2) for all other alpha emitters. (2) SCO-II: A solid object on which the limits for SCO-I are exceeded and on which:

(i) The non-fixed contamination on the accessible surface averaged over 300 cm^2 (or the area of the surface if less than 300 cm^2) does not exceed 400 Bq/cm^2 (10^{-2} microcurie/cm^2) for beta and gamma and low toxicity alpha emitters or 40 Bq/cm^2 (10^{-3} microcurie/cm^2) for all other alpha emitters;

(ii) The fixed contamination on the accessible surface averaged over 300 cm^2 (or the area of the surface if less than 300 cm^2) does not exceed 8×10^5 Bq/cm^2 (20 microcuries/cm^2) for beta and gamma and low toxicity alpha emitters, or 8×10^4 Bq/cm^2 (2 microcuries/cm^2) for all other alpha emitters; and

(iii) The non-fixed contamination plus the fixed contamination on the inaccessible surface averaged over 300 cm^2 (or the area of the surface if less than 300 cm^2) does not exceed 8×10^5 Bq/cm^2 (20 microcuries/cm^2) for beta and gamma and low toxicity alpha emitters, or 8×10^4 Bq/cm^2 (2 microcuries/cm^2) for all other alpha emitters.

Transport index	the dimensionless number (rounded up to the next tenth) placed on the label of a package, to designate the degree of control to be exercised by the carrier during transportation. The transport index is determined as follows: (1) For non-fissile material packages, the number determined by multiplying the maximum radiation level in millisievert (mSv) per hour at one meter (3.3 ft) from the external surface of the package by 100 (equivalent to the maximum radiation level in millirem per hour at one meter (3.3 ft)); or (2) For fissile material packages, the number determined by multiplying the maximum radiation level in millisievert per hour at one meter (3.3 ft) from the external surface of the package by 100 (equivalent to the maximum radiation level in millirem per hour at one meter (3.3 ft)), or, for criticality control purposes, the number obtained as described in 10 CFR 71.59, whichever is larger.
Type A quantity	a quantity of radioactive material, the aggregate radioactivity of which does not exceed A_1 for special form radioactive material, or A_2, for normal form radioactive material, where A_1 and A_2 are given in Table A-1 of 10 CFR Part 71, or may be determined by procedures described in Appendix A of 10 CFR Part 71.
Type B package	a Type B packaging together with its radioactive contents. On approval, a Type B package design is designated by NRC as B(U) unless the package has a maximum normal operating pressure of more than 700 kPa (100 lb/in^2) gauge or a pressure relief device that would allow the release of radioactive material to the environment under the tests specified in 10 CFR 71.73 (hypothetical accident conditions), in which case it will receive a designation B(M). B(U) refers to the need for

unilateral approval of international shipments; B(M) refers to the need for multilateral approval of international shipments. There is no distinction made in how packages with these designations may be used in domestic transportation. To determine their distinction for international transportation, see DOT regulations in 49 CFR Part 173. A Type B package approved before September 6, 1983, was designated only as Type B. Limitations on its use are specified in 10 CFR 71.13.

Type B quantity   a quantity of radioactive material greater than a Type A quantity.

U.S. Code of Federal Regulations (CFR)   organized by titles (e.g., Title 10, "Energy"), chapters (e.g., Chapter I, "U.S. Nuclear Regulatory Commission"), parts (e.g., Part 71, "Packaging and Transportation of Radioactive Material"), subparts (e.g., Subpart D, "Application for Package Approval"), and sections (e.g., 10 CFR 71.31). See also Title 49, "Transportation."

INTRODUCTION

The Standard Review Plan for Transportation Packages for Spent Nuclear Fuel, referred to here as the Standard Review Plan (SRP), provides guidance for the U.S. Nuclear Regulatory Commission (NRC) safety reviews of packages used in the transport of spent nuclear fuel (SNF) under Title 10 of the U.S. Code of Federal Regulations (CFR), Part 71 (10 CFR Part 71). It is not intended as an interpretation of NRC regulations. This SRP supplements NRC Regulatory Guide (RG) 7.9, "Standard Format and Content of Part 71 Applications for Approval of Packaging for Radioactive Material," for review of package applications. Nothing contained in this plan may be construed as having the force and effect of NRC regulations (except where the regulations are cited), or as indicating that applications supported by safety analyses and prepared in accordance with RG 7.9 will necessarily be approved, or as relieving any person from the requirements of 10 CFR Parts 20, 30, 40, 60, 70, or 71 or any other pertinent regulations. The principal purpose of the SRP is to ensure the quality and uniformity of staff reviews. It is also the intent of this plan to make information about regulatory matters widely available and improve communications between NRC, interested members of the public, and the nuclear industry, thereby increasing the understanding of NRC staff review process. In particular, this guidance assists potential applicants by indicating one or more acceptable means of demonstrating compliance with the applicable regulations.

The SRP is intended for use by NRC staff reviewers of package applications, amendments, and renewals. The SRP provides specific guidance for the staff's preparation of NRC safety evaluation report (SER). The SRP provides guidance relating to compliance with 10 CFR Part 71, and portions of other CFR titles and parts incorporated by reference in or applicable to 10 CFR Part 71.

The SRP is organized to correlate with the recommended content for a safety analysis report (SAR) as detailed in RG 7.9. The individual sections address the matters that are reviewed, the basis for the review, how the review is accomplished, the conclusions that are sought, and follow a common outline of subsections, illustrated below.

Appendix A provides a correlation of the SRP with 10 CFR Part 71 and RG 7.9.

Current packages for shipment of SNF are generally intended to be shipped only on an exclusive-use vehicle. NRC staff anticipates that future transport of SNF will also be made primarily by exclusive-use vehicle. Therefore, this SRP addresses only the regulatory requirements and acceptance criteria for exclusive-use shipment of SNF.

Subsection 1. Review Objective

> This subsection states the purpose and scope of the review of the SAR section in question.

Subsection 2. Areas of Review

> This subsection provides the general outline used for subsections 3, 4, 5 and 6 (see below). This subsection identifies the systems, components, analyses, data or other information that are reviewed as part of the particular SAR section in question.

Subsection 3. Regulatory Requirements

This subsection summarizes the requirements of 10 CFR Part 71 that relate to the SAR section in question. The requirements are organized in accordance with the major areas of review identified in Subsection 2 above.

Subsection 4. Acceptance Criteria

This subsection includes the regulatory requirements by reference and identifies other criteria that are acceptable practice for demonstrating that the package design meets the regulatory requirements. The criteria are organized in accordance with the major areas of review identified in Subsection 2 above.

This subsection typically identifies minimum acceptance criteria that are acceptable to the staff in dealing with a specific safety or design issue. These acceptance criteria are identified in the SRP so that staff reviewers can take uniform and well-understood positions as similar safety issues arise in future cases. Like RGs, these acceptance criteria are acceptable to the staff, but they are not considered as the only possible means of demonstrating compliance with applicable regulations.

Subsection 5. Review Procedures

This subsection provides guidance specifically developed for the reviewer in preparation of the SER. The review is organized in accordance with the areas of review identified in Subsection 2 above. Subsection 5 addresses procedures that the reviewer is to follow to provide verification that the applicable safety criteria have been met. In addition, it supplements the general requirement for review of all submitted documentation with guidance based on prior staff reviews, and NRC experience gained from the regulation of existing transportation packages.

To assist the reviewer, a chart is provided for the SAR section in question depicting the flow of pertinent information into, within, and from the review process.

Subsection 6. Evaluation Findings

This subsection provides examples of review conclusions appropriate for the SER. The findings are organized in accordance with the major areas of review identified in Subsection 2 above.

Subsection 7. References

This subsection identifies references used in review of the SAR section in question.

The Director of the Spent Fuel Project Office will direct and approve revisions, including clarifications, corrections, and modifications, as necessary.

Suggested revisions and other comments will be considered and should be sent to the Director, Spent Fuel Project Office, Office of Nuclear Material Safety and Safeguards, U.S. Nuclear Regulatory Commission, Washington, D.C. 20555-0001.

1 GENERAL INFORMATION REVIEW

1.1 REVIEW OBJECTIVE

The objective of this review is to establish (1) that the application includes an overview of relevant package information including intended use; and (2) a summary description of the packaging, operational features, and contents that provide reasonable assurance that the package can meet the regulations and operating objectives.

1.2 AREAS OF REVIEW

The SAR should be reviewed for adequacy of the package description and drawings of the packaging. Areas of review include the following:

1.2.1 General SAR Format

1.2.2 Package Design Information
1.2.2.1 Purpose of Application
1.2.2.2 Quality Assurance Program
1.2.2.3 Proposed Use/General Contents
1.2.2.4 Package Type and Model Number
1.2.2.5 Package Category and Maximum Activity
1.2.2.6 Materials Specifications, Fabrication, and Welding Criteria
1.2.2.7 Transport Index and Maximum Number of Packages

1.2.3 Package Description
1.2.3.1 Packaging
1.2.3.2 Operational Features
1.2.3.3 Contents of Packaging

1.2.4 Compliance with 10 CFR Part 71
1.2.4.1 General Requirements of 10 CFR 71.43
1.2.4.2 Condition of Package after Tests in 10 CFR 71.71 and 10 CFR 71.73
1.2.4.3 Structural, Thermal, Containment, Shielding, Criticality
1.2.4.4 Operational Procedures, Acceptance Tests and Maintenance

1.2.5 Appendix

1.3 REGULATORY REQUIREMENTS

Regulatory requirements of 10 CFR Part 71 applicable to the general information review are as follows:

1.3.1 General SAR Format

There are no specific regulatory requirements on the format of the SAR. SAR format provisions are given in RG 7.9.

1.3.2 Package Design Information

The application for package approval must: (1) include the classification and model number [10 CFR 71.31(a)(1), 10 CFR 71.33(a)(1), and 10 CFR 71.33(a)(3)]; (2) include a quality assurance program description or a reference to a previously approved quality assurance program applicable to the package [10 CFR 71.31(a)(3) and 10 CFR 71.37]; (3) identify applicable codes and standards used in package design, fabrication, assembly, testing, maintenance, and use [10 CFR 71.31(c)]; and (4) include the transport index for nuclear criticality control. [10 CFR 71.31(a)(2), 10 CFR 71.35(b), and 10 CFR 71.59]

An application for renewal of a previously approved package design must be submitted to NRC no later than 30 days prior to the expiration date of the approval to assure continued use and is subject to the provisions of 10 CFR 71.13. [10 CFR 71.38]

All changes in the conditions specified in the package approval must be approved by NRC. An application for modification of a previously approved package design may be subject to the provisions of 10 CFR 71.13(c) and 10 CFR 71.31(b). [10 CFR 71.107(c)]

1.3.3 Package Description

The description of the packaging must include a containment system, materials of construction, weights, dimensions, methods of fabrication, and coolant receptacle volumes in sufficient detail to provide an adequate basis for their evaluation. [10 CFR 71.31(a)(1), 10 CFR 71.33(a)(2), 10 CFR 71.33(a)(4), 10 CFR 71.33(a)(5), and 10 CFR 71.33(a)(6)]

The SAR must identify, with respect to the contents of the package, the maximum radioactive and fissile constituents, physical and chemical form, neutron absorbers or moderators, extent of reflection, moderator-to-fissile ratio, maximum normal operating pressure, maximum weight, maximum decay heat, and any coolant volumes. [10 CFR 71.31(a)(1) and 10 CFR 71.33(b)]

The outside of the package must incorporate a feature that, while intact, would be evidence that the package has not been opened by unauthorized persons. [10 CFR 71.43(b)]

Spent fuel, with plutonium in excess of 0.74 TBq (20 Ci) per package, in the form of debris, particles, loose pellets, or fragmented rods or assemblies must be packaged in a separate inner container (second containment system) in accordance with 10 CFR 71.63(b). [10 CFR 71.63]

1.3.4 Compliance with 10 CFR Part 71

The package must be evaluated to demonstrate compliance with the requirements specified in 10 CFR Part 71, Subpart E, under the conditions and tests of Subpart F. [10 CFR 71.31(a)(2), 10 CFR 71.35(a), and 10 CFR 71.41(a)]

1.4 ACCEPTANCE CRITERIA

1.4.1 General SAR Format

The application should be prepared in accordance with the general format provisions of RG 7.9.

1.4.2 Package Design Information

The regulatory requirements in Section 1.3.2 identify the acceptance criteria.

1.4.3 Package Description

In addition to the regulatory requirements identified in Section 1.3.3, a discussion of the operation of the package should be provided. [RG 7.9 (1.2.2)]

In addition to the regulatory requirements identified in Section 1.3.3, spent nuclear fuel with known or suspected cladding defects greater than a hairline crack or a pinhole leak should be canned. Canning of damaged fuel is to facilitate handling and to confine gross fuel particles to a known subcritical volume under normal conditions of transport and hypothetical accident conditions.

In addition to the regulatory requirements identified in Section 1.3.3, engineering drawings of the package should be provided. [RG 7.9 (1.3)]

1.4.4 Compliance with 10 CFR Part 71

In addition to the regulatory requirements identified in Section 1.3.4, a concise statement by the applicant, that the package complies with the requirements in 10 CFR Part 71 for a Type B(U)F package, should be provided in the General Information section of the SAR. This summary statement should provide a reference to the sections of the SAR that are used to specifically address compliance with the requirements of Subparts E and F of 10 CFR Part 71.

1.5 REVIEW PROCEDURES

The review should ensure that the General Information section of the SAR provides an adequate description of the spent nuclear fuel (SNF) transportation package so that its design and operation can be evaluated in subsequent sections. Although the General Information section of the SAR will not contain enough information by itself to perform a technical review of the package, the General Information section serves as a vehicle to facilitate consistency and reduce repetition between the various review disciplines (e.g., structural and shielding reviews), and presents summary information for the non-technical reviewers. The following procedures are generally applicable to the general information review of all SNF transportation packages.

Packages for shipment of SNF are generally intended to be shipped only on an exclusive-use vehicle. NRC staff anticipates that future transport of SNF will also be made primarily by exclusive-use vehicle.

Therefore, this SRP addresses only the regulatory requirements and acceptance criteria for exclusive-use shipment of SNF.

The general information review is based in part on the descriptions and evaluations presented in the General Information section of the SAR and follows the sequence established to evaluate the packaging against applicable 10 CFR Part 71 requirements. Similarly, results of the general information review are considered in the review of the SAR sections on Structural Evaluation, Thermal Evaluation, Containment Evaluation, Shielding Evaluation, Criticality Evaluation, Operating Procedures, and Acceptance Tests and Maintenance Program. Examples of SAR information flow within and from the general information review are shown in Figure 1-1.

1.5.1 General SAR Format

Verify that the SAR has been prepared in accordance with the use of standard format, style and composition, revisions, and physical specifications described in RG 7.9 (i.e., paper size, paper stock, ink, page margins, printing, binding, page numbering, separators, and number of copies).

1.5.2 Package Design Information

1.5.2.1 Purpose of Application

The purpose of the application should be clearly stated. The application may be for approval of a new design, for amendment, or for renewal of an existing approval (i.e., certificate of compliance). Applications for approval of a new design should be whole and complete and should contain the information identified in Subpart D of 10 CFR Part 71. If the application is for modification of an approved design, verify that the changes being requested are clearly identified. Modifications may include design changes, changes in authorized contents, or changes in conditions of the approval. Design changes should be clearly identified and should be included in revised packaging drawings. Packagings that do not conform to the drawings referenced in the NRC approval are not authorized for use under the general license in 10 CFR 71.12. Likewise, only contents specified in the approval may be transported. Package operating procedures, acceptance tests, and a maintenance program may also be specified as conditions of the approval.

Applications for modifications to an approved design should include an assessment of the requested changes and justification that these changes do not affect the ability of the package to meet the requirements of 10 CFR Part 71. Applications for modifications may be subject to the provisions of 10 CFR 71.13(c) and 10 CFR 71.31(b), as applicable. When the modification is submitted under the provision of 10 CFR 71.13(c)(1) or 10 CFR 71.13(c)(2), the application should justify that the requested change is not significant.

Applications for renewal of an existing approval should be made not less than 30 days before the expiration of the approval to assure continued use. Applications for renewal are subject to the provisions of 10 CFR 71.38.

1.5.2.2 Quality Assurance Program

Verify that the applicant has obtained NRC approval of its quality assurance program, or has identified by reference a quality assurance program that has been previously approved under the requirements of 10 CFR 71.12, 10 CFR 71.37 and 10 CFR Part 71, Subpart H.

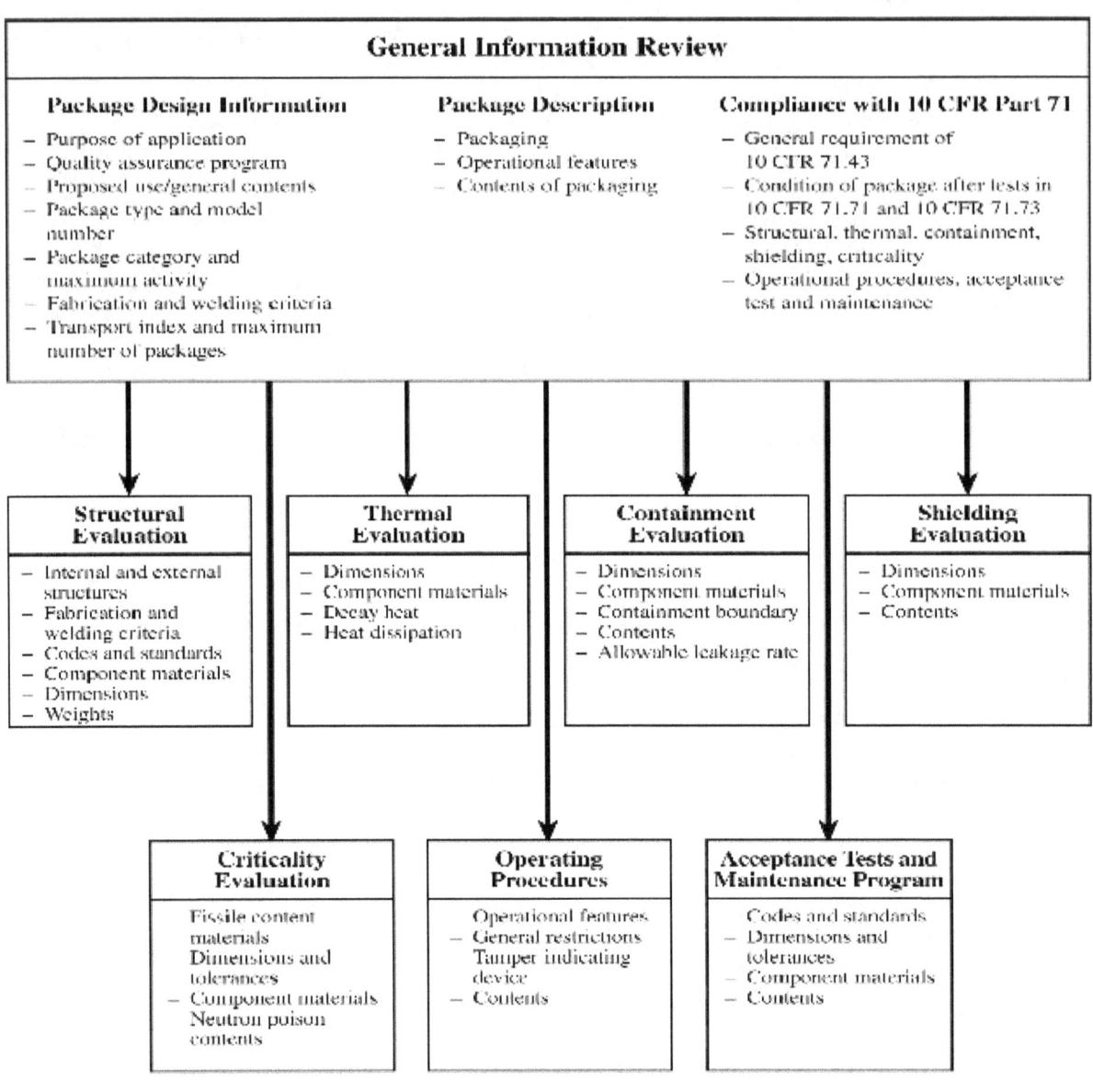

General Information Review

Package Design Information	Package Description	Compliance with 10 CFR Part 71
– Purpose of application – Quality assurance program – Proposed use/general contents – Package type and model number – Package category and maximum activity – Fabrication and welding criteria – Transport index and maximum number of packages	– Packaging – Operational features – Contents of packaging	– General requirement of 10 CFR 71.43 – Condition of package after tests in 10 CFR 71.71 and 10 CFR 71.73 – Structural, thermal, containment, shielding, criticality – Operational procedures, acceptance test and maintenance

Structural Evaluation

– Internal and external structures
– Fabrication and welding criteria
– Codes and standards
– Component materials
– Dimensions
– Weights

Thermal Evaluation

– Dimensions
– Component materials
– Decay heat
– Heat dissipation

Containment Evaluation

– Dimensions
– Component materials
– Containment boundary
– Contents
– Allowable leakage rate

Shielding Evaluation

– Dimensions
– Component materials
– Contents

Criticality Evaluation

Fissile content materials
Dimensions and tolerances
– Component materials
Neutron poison contents

Operating Procedures

Operational features
– General restrictions
Tamper indicating device
– Contents

Acceptance Tests and Maintenance Program

Codes and standards
– Dimensions and tolerances
– Component materials
– Contents

Figure 1-1 SAR Information Flow for the General Information Review

1.5.2.3 Proposed Use/General Contents

Verify that the description for the proposed use of the packaging and the general contents of the package are sufficient to allow the reviewer to understand exactly how the packaging is to be used and what is to be transported. Since packages for shipment of SNF are generally intended to be shipped by exclusive use, only exclusive-use shipments are assumed in the following SRP review procedures. Verify that the package is to be shipped by exclusive use and ensure that any restrictions regarding the use or type of conveyance are designated.

1.5.2.4 Package Type and Model Number

Confirm that the type and model number of the package are designated. A new SNF transportation package will be designated B(U)F-85 unless it has a maximum normal operating pressure (MNOP) greater than 700 kPa (100 lb/in^2) or a pressure relief device that would allow the release of radioactive material under the tests specified in 10 CFR 71.73 (hypothetical accident conditions). In those cases, the package will be designated B(M)-85. Verify that a model number is designated for the package and that it is specified on the appropriate drawings.

1.5.2.5 Package Category and Maximum Activity

Category I is assigned to a package whose content activity exceeds either 1.11×10^{15} Bq (30,000 Ci), 3000 A_1, or 3000 A_2 whichever is less. (SNF transportation packages are assumed to be Category I in the following SRP review procedures.) Verify that the package is designated Category I and that the maximum activity of the package contents is specified.

1.5.2.6 Materials Specifications, Fabrication, and Welding Criteria

ASME has published Section III, Division 3, ASME Boiler and Pressure Vessel (B&PV Division 3) Code for the design and construction of the containment system of SNF transport packagings. NRC staff expects full compliance with the B&PV Division 3 Code for the containment system, including the services of an Authorized Inspection Agency. However, the SAR may justify alternatives as appropriate. The code used for the design should also be used for materials specifications, fabrication, and welding criteria.

Criteria acceptable for other components of SNF transportation packages are also based on the ASME Boiler and Pressure Vessel (B&PV) Code. Table 1-1 summarizes the appropriate B&PV Code sections for the fabrication, examination, and testing of the Containment, Criticality, and Other Safety component groups. Table 1-2 summarizes the appropriate B&PV Code sections for welding of the key elements of the Containment, Criticality, and Other Safety component groups.

Verify that the fabrication, welding, and examination criteria for the package are specified for each major component and that they are appropriate for a SNF package. Verify that materials specifications and standards have been specified for all major components and that they are consistent for the product form to be fabricated. Verify that a reference is provided to the sections of the SAR where a discussion of any fabrication, welding, and examination of package components may be found.

Table 1-1 Fabrication, Examination, and Testing Criteria for SNF Transportation Packages based on the B&PV Code.[1]

Container contents	Containment		Criticality	Other Safety							
	Primary vessel, bolts, piping, fittings, valves, closure	Primary seal	Support structures/ neutron absorber[2]	Gamma shielding[3]	Structural shell, bolts, and closure	Secondary seal	Neutron shielding, piping, fittings, valves, relief device, and tanks[4]	Lifting lugs	Impact limiters[5]	Tie down devices	Heat transfer devices
B&PV Code section	Sec. III, Division 3		Sec. III, Subsection NG	Sec. VIII, Div. 1[6] or Sec. III, Subsection NF							
Materials[7]	WB-2000	8	NG-2000								
Forming, fitting and aligning	WB-4200[3]		NG-4200								
Heat treatment	WB-4600		NG-4600								
Examination	WB-5000		NG-5000								
Acceptance testing	WB-6000[9]	9				8		10			11

1   These criteria should be referenced in the associated SAR. Criteria for special processes used, but not included in this table, should be documented in the SAR. Fabrication criteria for welding and brazing are recommended in Table 1-2. Quality assurance criteria are provided in 10 CFR Part 71 and RG 7.10. Referenced supporting portions of Section II; Section III, Subsection NCA; Section V; and Section IX of the B&PV Code are part of the recommended criteria.

2   The designer may specify a neutron absorber material by a commercial trade name or as a mixture of elements or common compounds. When appropriate, qualification data should be included in the SAR to demonstrate that the material functions as specified. When special absorber materials are used to control criticality, an acceptance test should be performed for each container to ensure that the absorber material has been properly installed. Structural criteria do not apply to neutron absorbers unless they are used for structural support.

3   The installation of shielding may involve processes such as lead pouring around the primary vessel or shrink fitting of uranium castings onto the primary vessel which could affect the vessels structural integrity. In such cases, the fabrication criteria for the specific process and an engineering evaluation of any associated structural effects should be performed to ensure its effectiveness.

4. The designer may specify a neutron shielding material by a commercial trade name or as a mixture of elements or common compounds. When appropriate, qualification data should be included in the SAR to demonstrate that the material functions as specified. Acceptance testing may be required to demonstrate the effectiveness of the neutron shielding.

5. Impact limiters may use special materials such as wood or honeycomb metals to provide the specified crushing characteristics. Any special processes, physical properties, or other information needed to install the impact limiter or qualify its proper function should be included in the SAR.

6. Specific articles in Section VIII, Division 1 have not been listed since the fabrication process is dependent on the fabrication method and materials used. Once the method of fabrication and materials of construction have been specified, the appropriate fabrication criteria can be found in Subsection A, General Requirements; Subsection B, Methods of Fabrication; and Subsection C, Classes of Materials. Criteria from equivalent ASTM materials and standards, DOT specifications or articles in B&PV Code Section III, Subsection NF may also be substituted, all or in part.

7. The B&PV Code was written for pressure vessel and nuclear component fabrication and does not include many of the materials used in the shipping container industry. The designer may specify the material to be used by either a commercial trade name or an applicable ASTM specification. For each material used, information or references should be included in the SAR to permit an evaluation of the materials properties and the intended use. For thicknesses up to four inches, ferritic materials should satisfy the fracture toughness criteria recommended in RG 7.11 for the relevant container category instead of the fracture toughness criteria specified in the B&PV Code. Fracture toughness criteria for ferritic steel thicknesses greater than four inches are in RG 7.12.

8. The B&PV Code does not have specifications for either gasket or seal materials. The designer may specify the material and configuration by a commercial trade name. Information which demonstrates the qualification of the seal or gasket (including those used for valves and relief devices) should be included in the SAR.

9. Leak testing of the primary containment, including seals, should be performed in accordance with ANSI N14.5.

10. Shipping containers involved in critical lifts in nuclear facilities should have their lifting lugs fabricated and tested to the criteria specified in NUREG-0612 and ANSI N14.6.

11. Heat transfer devices required to contain pressure should be hydrostatically tested to Section VIII, Subsection UG-99. Acceptance testing for each shipping container may be necessary to ensure that the specified heat transfer rate is obtained.

Table 1-2 Welding Criteria for SNF Transportation Packages.

Key welding elements	Weld type		
	Containment-related	**Criticality-related**	**Other safety-related**
B&PV Code section	Sec. III, Division 3	Sec. III, Subsection NG	Sec. VIII, Div. 1 or Sec. III, Subsection NF
Base materials	WB-2000, WB-4100, and applicable Code cases	NG-2100, NG-2200, NG-2500, NG-4100, and applicable Code cases	Sec. VIII, Div. 1, Subsection A, General Requirements; appropriate parts of Subsection B, Methods of Fabrication; and Subsection C, Classes of Materials. Or Sec. III, Subsection NF
Welding and brazing materials	WB-2400	NG-2400	
Joint preparation	WB-4200	NG-4200	
Welding	WB-4400	NG-4400	
Brazing	WB-4500	NG-4500	
Heat treatment	WB-4600	NG-4600	
Qualification of procedures and personnel	WB-4300	NG-4300	
Examination	WB-5000	NG-5000	
Quality assurance	10 CFR Part 71, Subpart H and RG 7.10		
Fracture toughness	RG 7.11 or 7.12	As justified in the SAR	

1.5.2.7 Transport Index and Maximum Number of Packages

Verify that a transport index has been assigned to the packaging for the SNF contents and that a reference is provided to the section of the SAR where a discussion of the determination of the transport index is found.

Verify that the maximum number of SNF packages in one shipment has been assigned to the packaging for the specified fissile contents and that a reference is provided to the section of the SAR where the determination of the maximum number of packages is found.

1.5.3 Package Description

1.5.3.1 Packaging

Review the text description of the packaging and verify that the following information, as applicable, is discussed. Sketches, figures, or other schematic diagrams should be used as appropriate.

• The gross weight, external dimensions, and cavity size

- Materials of construction, weights, dimensions, and fabrication methods of receptacles, neutron absorbers or moderators, internal and external structures supporting or protecting receptacles, fuel basket and engineered flux traps, valves, sampling ports, lifting and tie-down devices, impact limiters, structural and mechanical means of heat dissipation, types of coolant, outer and inner protrusions, shielding, pressure relief systems, and closures

- Identification of the containment system and boundary (see Section 4.5.1.3 for additional guidance on containment of damaged fuel).

Examine the detailed drawings presented in the appendix. Verify that information shown on the drawings is consistent with that discussed in the text. Drawings should be sufficiently detailed to provide a package description that can be evaluated for compliance with 10 CFR Part 71. The packaging drawings are incorporated by reference into the certificate of compliance and become regulatory conditions for compliance.

Confirm that each drawing has a title block that identifies the preparing organization, drawing number, sheet number, title, date, and signature or initials indicating approval of the drawing. Revised drawings should identify the revision number, date, and description of the change in each revision. Proprietary information should be clearly identified. The drawings should include:

- General arrangement of the packaging and contents, including dimensions

- Design features which affect the package evaluation

- Package markings

- Maximum allowable weight of the package

- Maximum weight of contents and secondary packaging

- Minimum weights, if appropriate.

Information on design features should include, as appropriate:

- Identification of the design feature and its components

- Materials of construction, including appropriate material specifications

- Codes, standards, or other similar specification documents for fabrication, assembly, and testing

- Dimensions with appropriate tolerances

- Operational specifications (e.g., bolt torque)

- Tamper indicating device.

Additional guidance on engineering drawings submitted in the SAR is provided in NUREG/CR-5502.

1.5.3.2 Operational Features

For complex packages, verify that all operational features and functions are discussed. A schematic diagram should be included in the SAR showing all valves, connections, piping, openings, seals, and containment boundaries. Detailed operational schematics should be provided and annotated in accordance with the operations described in the Operating Procedures section of the SAR. However, details may be referenced in the General Information section of the SAR, if provided in a later SAR section or appendix. In this case, simplified operational schematics should be an acceptable alternative in the General Information section of the SAR. Loading configurations for all contents should be provided and annotated in a manner consistent with the Structural Evaluation, containment Evaluation, Thermal Evaluation, Shielding Evaluation, Criticality Evaluation, and Operating Procedures sections of the SAR. Confirm that a reference is provided to any other section of the SAR where evaluations of the operability and safety of the operational features are found.

Any codes and standards proposed for regulating the operation of the package should be identified and a reference provided to any other section of the SAR where a discussion of the proposed codes and standards is found. Confirm that a reference is provided.

1.5.3.3 Contents

The contents should be described in the same detail as that intended for the certificate of compliance. Review the description of the contents and verify that, as a minimum, the following information is presented:

- The type of SNF and maximum initial U-235 mass, its associated burnup, specific power, cooling time, heat load, and maximum and minimum initial enrichment, including a description of non-uniform enrichment, if applicable

- Fuel assembly specifications, including dimensional data for the fuel rods and assembly structure

- Control assemblies or other contents (e.g., startup sources) that may be present

- Maximum quantities of radionuclides present in the SNF and the quantities estimated to be available for immediate release within the void space of the fuel rods

- Maximum quantity of unirradiated fuel and maximum initial U-235 mass per assembly or rods and number of assemblies or rods

- Chemical and physical form, presence of any annular pellets

- Location and configuration within the packaging

- Any material subject to chemical, galvanic, or other reaction, including the generation of combustible gases

- Fuel densities

- Amounts of neutron absorbers or moderators in the fuel or package

- Basket or other configurations of fuel assemblies or rods

- MNOP

- Maximum weight

- Free volume of the containment vessel

- Containment fill gas

- Any unique or unusual conditions (e.g., failed fuel, non-uniform enrichment, etc.)

- For damaged fuel, the maximum quantity of damaged fuel, initial enrichment, absorption, extent of damage, and description of the second containment system, and any other limits, as applicable are specified (see also Section 4.5.1.3).

1.5.4 Compliance with 10 CFR Part 71

Review the summary results to determine if the packaging complies with regulations.

1.5.4.1 General Requirements of 10 CFR 71.43

Verify that a summary statement is provided indicating compliance with the general standards for all packages and that references are provided to the sections of the SAR where discussions of compliance with the general standards for all packages are found.

1.5.4.2 Condition of Package after Tests in 10 CFR 71.71 and 10 CFR 71.73

Verify that summary descriptions are provided for the physical condition of the package subsequent to the tests specified in 10 CFR 71.71 (normal conditions of transport) and 10 CFR 71.73 (hypothetical accident conditions). Verify that references are provided to all sections of the SAR where discussions of the physical conditions of the package subsequent to testing are found.

1.5.4.3 Structural, Thermal, Containment, Shielding, Criticality

Verify that summary statements are provided attesting to the adequacy of the package design to meet the structural, thermal, containment, shielding, and criticality requirements of 10 CFR Part 71.

1.5.4.4 Operational Procedures, Acceptance Tests and Maintenance

Verify that a summary statement is provided attesting to the adequacy of the development of the operational procedures and acceptance tests and maintenance program to ensure compliance with the requirements of 10 CFR Part 71.

1.5.5 Appendix

In addition to the packaging drawings discussed above, the appendix may include a list of references and copies of any applicable references not generally available to the reviewer. The appendix may also provide supporting details on special fabrication procedures and other appropriate supplemental information.

1.6 EVALUATION FINDINGS

The Safety Evaluation Report (SER) does not normally include specific findings for the General Information section of the SAR.

Before proceeding with the review of the other sections of the SAR, the reviewer should conclude, at a minimum, that the following criteria have been demonstrated:

- The package has been described in sufficient detail to provide an adequate basis for its evaluation.

- Drawings provided contain information which provides an adequate basis for its evaluation against 10 CFR Part 71 requirements. Each drawing is identified, consistent with the text of the SAR, and contains keys or annotation to explain and clarify information on the drawing.

- The application for package approval includes a reference to the approved quality assurance program for the package.

- The application for package approval identifies applicable codes and standards for the package design, fabrication, assembly, testing, maintenance, and use.

- The package meets the general requirements of 10 CFR 71.43(a) and 10 CFR 71.43(b).

- Drawings submitted with the application provide a detailed packaging description that can be evaluated for compliance with 10 CFR Part 71 for each of the technical disciplines.

- Any restrictions on the use of the package are specified.

- Any modifications to a previously approved package do not violate the restrictions in 10 CFR 71.13(c).

1.7 REFERENCES

ANSI N14.5   Institute for Nuclear Materials Management, ANSI N14.5, "Leakage Tests on Packages for Shipment of Radioactive Materials," New York, NY, 1987.

ANSI N14.6   Institute for Nuclear Materials Management, ANSI N14.6, "Special Lifting Devices for Shipping Containers Weighing 10,000 Pounds (45000 kg) or More for Nuclear Materials," New York, NY, 1993.

B&PV Code	American Society of Mechanical Engineers, "ASME Boiler and Pressure Vessel Code," New York, NY, 1998.
B&PV Division 3 Code	American Society of Mechanical Engineers, "ASME Boiler and Pressure Vessel Code, Section III, Division 3, Containment Systems and Transport Packagings For Spent Nuclear Fuel and High Level Radioactive Waste," New York, NY, 1998.
NUREG-0612	U.S. Nuclear Regulatory Commission, "Control of Heavy Loads at Nuclear Power Plants," NUREG-0612, National Technical Information Service, Springfield, VA, July 1980.
NUREG/CR-5502	U.S. Nuclear Regulatory Commission, "Engineering Drawings for 10 CFR Part 71 Package Approvals," NUREG/CR-5502, U.S. Government Printing Office, Washington, D.C., May 1998.
RG 7.9	U.S. Nuclear Regulatory Commission, Regulatory Guide 7.9, "Standard Format and Content of 71 Applications for Approval of Packaging of type B, Large Quantity and Fissile Radioactive Material," U.S. Government Printing Office, Washington, D.C., July 1979.
RG 7.10	U.S. Nuclear Regulatory Commission, Regulatory Guide 7.10, "Establishing Quality Assurance Programs for Packaging Used in the Transport of Radioactive Material," U.S. Government Printing Office, Washington, D.C., January 1983.
RG 7.11	U.S. Nuclear Regulatory Commission, Regulatory Guide 7.11, "Fracture Toughness Criteria of Base Material for Ferritic Steel Shipping Cask Containment Vessels with a Maximum Wall Thickness of 4 Inches (0.1 m)," U.S. Government Printing Office, Washington, D.C., June 1991.
RG 7.12	U.S. Nuclear Regulatory Commission, Regulatory Guide 7.12, "Fracture Toughness Criteria of Base Material for Ferritic Steel Shipping Cask Containment Vessels with a Wall Thickness Greater than 4 Inches (0.1 m)," U.S. Government Printing Office, Washington, D.C., June 1991.

2 STRUCTURAL REVIEW

2.1 REVIEW OBJECTIVE

The objective of this review is to verify that the structural performance of the package has been adequately evaluated for the tests specified under normal conditions of transport and hypothetical accident conditions and that the package design has adequate structural integrity to meet the requirements of 10 CFR Part 71.

2.2 AREAS OF REVIEW

The SAR should be reviewed for adequacy of the description and evaluation of the structural design. Areas of review include the following:

2.2.1 Description of Structural Design
2.2.1.1 Descriptive Information Including Weights and Centers of Gravity
2.2.1.2 Codes and Standards

2.2.2 Material Properties
2.2.2.1 Materials and Material Specifications
2.2.2.2 Chemical, Galvanic, or Other Reactions
2.2.2.3 Effects of Radiation on Materials

2.2.3 Lifting and Tie-down Standards for All Packages
2.2.3.1 Lifting Devices
2.2.3.2 Tie-down Devices

2.2.4 General Considerations for Structural Evaluation of Packaging
2.2.4.1 Evaluation by Analysis
2.2.4.2 Evaluation by Test

2.2.5 Normal Conditions of Transport
2.2.5.1 Heat
2.2.5.2 Cold
2.2.5.3 Reduced External Pressure
2.2.5.4 Increased External Pressure
2.2.5.5 Vibration
2.2.5.6 Water Spray
2.2.5.7 Free Drop
2.2.5.8 Corner Drop
2.2.5.9 Compression
2.2.5.10 Penetration

2.2.6 Hypothetical Accident Conditions
2.2.6.1 Free drop

2.2.6.2 Crush

2.2.6.3 Puncture

2.2.6.4 Thermal

2.2.6.5 Immersion – Fissile Material

2.2.6.6 Immersion – All Packages

2.2.7 Special Requirement for Irradiated Nuclear Fuel Shipments

2.2.8 Internal Pressure Test

2.2.9 Appendix

2.3 REGULATORY REQUIREMENTS

Regulatory requirements of 10 CFR Part 71 applicable to the structural review are as follows:

2.3.1 Description of Structural Design

The packaging must be described in sufficient detail to provide an adequate basis for its evaluation. [10 CFR 71.31(a)(1) and 10 CFR 71.33]

The SAR must identify established codes and standards applicable to the structural design and fabrication of the package. [10 CFR 71.31(c)]

2.3.2 Material Properties

The package must be made of materials which assure that there will be no significant chemical, galvanic, or other reactions among the packaging components, among package contents, or between the packaging components and the package contents, including possible reaction resulting from inleakage of water. The effects of radiation on the materials of construction must also be considered. [10 CFR 71.43(d)]

2.3.3 Lifting and Tie-down Standards for All Packages

The package design must meet the lifting and tie-down requirements of 10 CFR 71.45.

2.3.4 General Considerations for Structural Evaluation of Packaging

The package must be evaluated to demonstrate that it satisfies the standards specified in 10 CFR Part 71, Subpart E, under the conditions and tests of Subpart F. [10 CFR 71.31(a)(2), 10 CFR 71.35(a), and 10 CFR 71.41(a)]

The effects on the performance of the package under the tests specified in 10 CFR 71.71 (normal conditions of transport), in 10 CFR 71.73 (hypothetical accident conditions), and in 10 CFR 71.61 (special requirement for irradiated nuclear fuel shipments) must be evaluated. [10 CFR 71.41(a)] by subjecting a specimen or scale model to a specific test, or by another appropriate and acceptable method.

2.3.5 Normal Conditions of Transport

The package must be evaluated under the tests specified in 10 CFR 71.71 for normal conditions of transport. [10 CFR 71.41(a)] The evaluation must show that under the tests, there would be no substantial reduction in the effectiveness of the packaging. [10 CFR 71.35(a), 10 CFR 71.43(f), 10 CFR 71.51(a)(1), and 10 CFR 71.55(d)(4)]

2.3.6 Hypothetical Accident Conditions

The package must be evaluated under the tests specified in 10 CFR 71.73 for hypothetical accident conditions. [10 CFR 71.41(a)] The evaluation must show that the packaging has adequate structural integrity to satisfy the containment, shielding, subcriticality, and temperature requirements of 10 CFR Part 71, Subpart E. [10 CFR 71.35(a)]

2.3.7 Special Requirement for Irradiated Nuclear Fuel Shipments

The containment vessel of a package with activity greater than 37 PBq (10^6 Ci) must be designed to withstand an external pressure of 2 MPa (290 psi) for a period of not less than one hour without collapse, buckling, or inleakge of water. [10 CFR 71.61]

2.3.8 Internal Pressure Test

Where the maximum normal operating pressure (MNOP) will exceed 35 kPa (5 psi) gauge, the containment design of all packages must be tested at an internal pressure at least 50 percent higher than the MNOP to verify that the system can maintain structural integrity at that pressure. [10 CFR 71.85(b)]

2.4 ACCEPTANCE CRITERIA

2.4.1 Description of Structural Design

In addition to the regulatory requirements identified in Section 2.3.1, the containment system should be designed and constructed in accordance with Section III, Division 3, ASME Boiler and Pressure Vessel (B&PV Division 3) Code. Alternate codes should be justified in the SAR.

In addition to the regulatory requirements identified in Section 2.3.1, packaging components other than the containment vessel should be designed and constructed in accordance with the criteria identified in Section 1.5.2.6 and Section 2.5.1.2.

In addition to the regulatory requirements identified in Section 2.3.1, load combinations for the packaging design should be in accordance with RG 7.8.

2.4.2 Material Properties

In addition to the regulatory requirements identified in Section 2.3.2, the structural design should address precluding brittle fracture in containments made of ferritic steels as described in RGs 7.11 and 7.12.

Material properties should meet the material specifications applicable to the codes and standards used for the design and fabrication of the package.

2.4.3 Lifting and Tie-Down Standards for All Packages

Section 2.3.3 discusses the acceptance criteria for lifting and tie-down standards.

2.4.4 General Considerations for Structural Evaluation of Packaging

Section 2.3.4 discusses the structural design criteria.

2.4.5 Normal Conditions of Transport

Section 2.3.5 discusses the acceptance criteria for package testing with respect to normal transport conditions.

2.4.6 Hypothetical Accident Conditions

Section 2.3.6 discusses the acceptance criteria for package testing with respect to hypothetical accident conditions..

2.4.7 Special Requirement for Irradiated Nuclear Fuel Shipments

Section 2.3.7 discusses the acceptance criteria for package testing with activity greater than 37 PBq.

2.4.8 Internal Pressure Test

Section 2.3.8 discusses the acceptance criteria for internal pressure testing.

2.5 REVIEW PROCEDURES

The following procedures are generally applicable to the structural review of all spent nuclear fuel (SNF) transportation packages.

The structural review is based in part on the descriptions and evaluations presented in the General Information and Thermal Evaluation sections of the SAR. Similarly, results of the structural review are considered in the review of the SAR sections on Thermal Evaluation, Containment Evaluation, Shielding Evaluation, Criticality Evaluation, Operating Procedures, and Acceptance Tests and Maintenance Program. Examples of SAR information flow into, within, and from the structural review are shown in Figure 2-1.

2.5.1 Description of Structural Design

2.5.1.1 Descriptive Information Including Weights and Centers of Gravity

Review drawings and other descriptions of the structural design in the General Information and Structure Evaluation sections of the SAR. The information should describe the function, geometry, and material of construction of all structural components of the packaging and its lifting and tie-down devices. The information should be sufficient for evaluating the structural performance of the packaging to meet the regulatory requirements, which include containment, shielding, and maintaining subcriticality of the

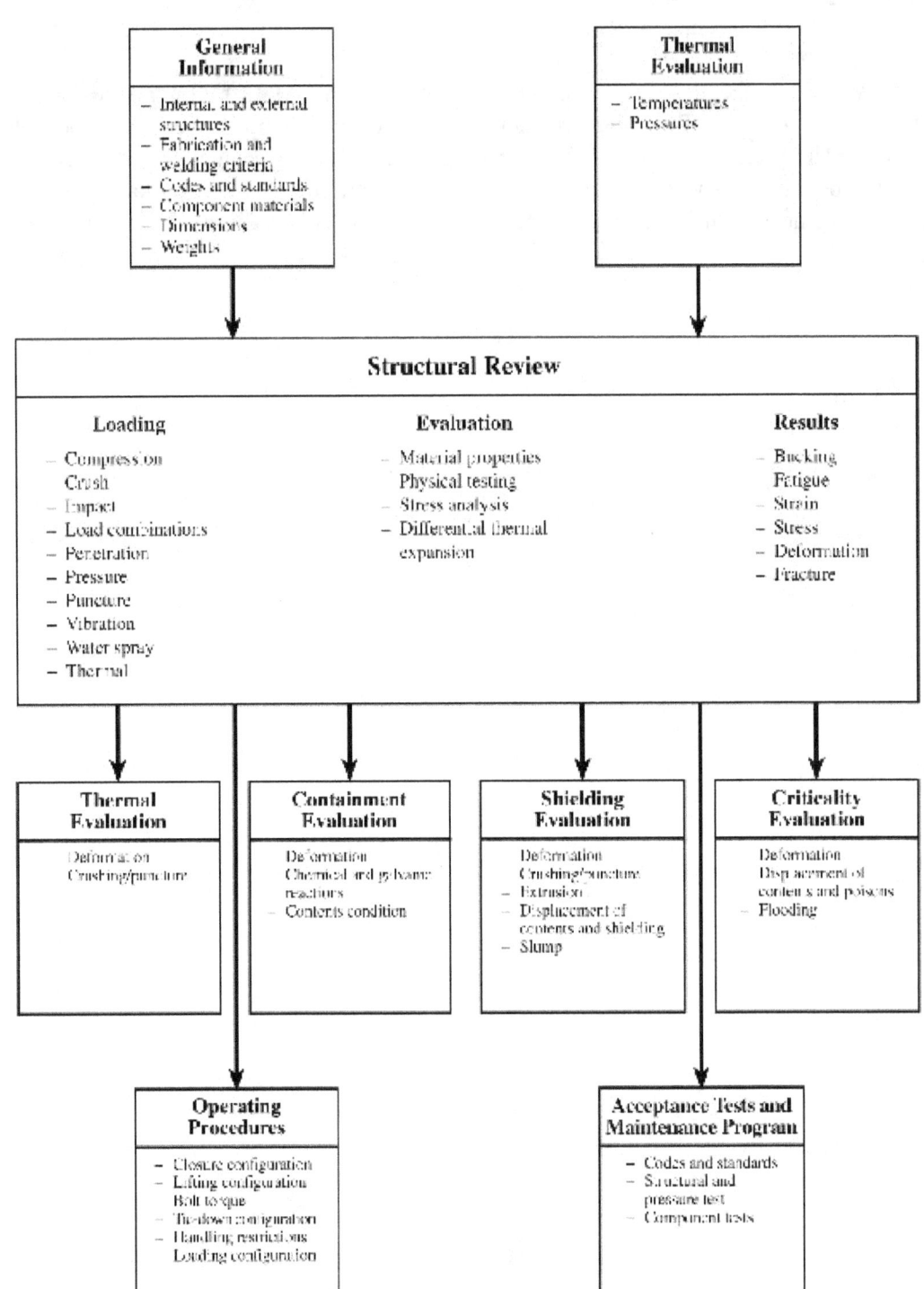

Figure 2-1 SAR Information Flow for Structural Review

radioactive contents under the normal conditions of transport and the hypothetical accident conditions. Verify that the data used in the structural evaluation are consistent with those on the drawings and descriptions of the structural design in the SAR.

Verify that packaging drawings provided in General Information section of the SAR specify the materials of construction, dimensions, tolerances and fabrication methods of the packaging and subassemblies, receptacles, internal or external support structures, valves and ports, lifting devices, and tie-down devices, and other design features relevant to the structural evaluation. Descriptive information such as the maximum and minimum weight of the package, the maximum weight of the contents, the center of gravity of the package, and the MNOP should be included.

2.5.1.2 Codes and Standards

The SAR should identify established codes and standards or justify the basis used for the package design and fabrication. The codes and standard must be appropriate for the intended purpose, and must be properly applied. The reviewer should verify that the code or standard:

- Was developed for structures of similar design and material, if not specifically for shipping packages

- Was developed for structures with similar loading conditions

- Was developed for structures which have similar consequences of failure

- Adequately addresses potential failure modes

- Adequately addresses margins of safety.

The ASME has developed a code specifically for the design and construction of the containment systems of an SNF or high-level radioactive waste transport packaging (B&PV Division 3 Code). NRC will accept the material, design, fabrication, welding, examination, testing, inspection, and certification of containment systems for SNF transportation packagings in accordance with the B&PV Division 3 Code. If there are any deviations in any way from the B&PV Division 3 Code, the SAR should explicitly state the applicant's justification for the deviation, and the justification must be acceptable to NRC.

NUREG/CR-3854 identifies codes and standards which may be used for fabricating components of SNF transportation packaging. Detailed recommendations of this report are summarized in Section 1.5.2.6, Table 1-1.

Several RGs and NUREGs provide guidance for structural design evaluation of packages using information from existing codes and practices: (1) RG 7.8 identifies the load combinations to be used in package design evaluation, (2) RG 7.6 provides design stress criteria for the containment system of Type B packages, (3) RGs 7.11 and 7.12 describe criteria for precluding brittle fracture in package containers made of ferritic steels, (4) NUREG/CR-4554A discusses the buckling evaluation of containment vessels, (5) NUREG/CR-6322 provides guidance for buckling analysis of SNF baskets, (6) NUREG/CR-6007 provides guidance and criteria for design analysis of closure bolts for packagings, and (7) NUREG/CR-3019 presents criteria for transportation package welds.

Guidance applicable for trunnions is provided in NUREG-0612 and ANSI N14.6

2.5.2 Material Properties

2.5.2.1 Materials and Material Specifications

Review packaging materials of construction and their specifications. Material specifications and properties should be consistent with the design code or standard selected; if no standard is available, the SAR should provide adequately documented material properties that are important for the design and fabrication of the packaging. A list of pertinent material properties needed to define the material for analysis should be provided.

Verify that the materials of structural components whose structural integrity is essential for the package to meet regulatory requirements have sufficient fracture toughness to preclude brittle fracture under the specified normal conditions of transport and hypothetical accident condition temperatures and loads. Brittle fracture must be precluded for the containment vessel under severe impact loads at the lowest service temperature. Fracture toughness criteria for ferritic steel packaging containment vessels are provided in RGs 7.11 and 7.12.

Verify that the material properties used are appropriate for the load condition (e.g., static or dynamic impact loading, hot or cold temperature, wet or dry conditions, etc.). Verify that appropriate temperatures at which allowable stress limits are defined are consistent with those temperatures expected in service and determined in the thermal analysis.

If the package has impact limiters, the adequacy of the method used for establishing their force-deflection characteristics should be verified by testing. Testing of the impact limiters may be carried out statically, if the effect of strain rates on the material crush properties is accounted for and properly included in the force-deflection relationship for impact analysis. The force-deflection curve of the impact limiter should be provided in the SAR for all directions evaluated for the packaging.

2.5.2.2 Prevention of Chemical, Galvanic, or Other Reactions

Review the materials and coatings of the package to verify that they will not produce a significant chemical or galvanic reaction among packaging components, among packaging contents, or between the packaging components and the packaging contents. The review should also include consideration of a possible reaction resulting from inleakage of water.

Evaluate the possible generation of hydrogen or other flammable gases; if appropriate, consider embrittling effects of hydrogen taking into account the metallurgical state of the packaging materials.

For metallic components of the package that may come into physical contact with one another, the possibility of eutectic reactions should be considered since such reactions can lead to melting at the interface between the metals at a lower temperature than the melting points of the metals in contact. Review methods used to prevent eutectic reactions.

2.5.2.3 Effects of Radiation on Materials

Verify that any damaging effects of radiation on the packaging materials have been appropriately considered. These effects include degradation of seals and sealing materials, and degradation of the properties of coatings and structural materials.

2.5.3 Lifting and Tie-Down Standards for All Packages

2.5.3.1 Lifting Devices

Review the design and evaluation of those lifting devices that are a structural part of the package, their connection with the package body, and the package body in the local area around the lifting devices. Verify that the design, testing, and analyses demonstrate that these devices comply with the requirements of 10 CFR 71.45(a):

- Any lifting attachment which is a structural part of the package must be designed with a minimum safety factor of three against yielding when used to lift the package in the intended manner

- A lifting attachment which is a structural part of the package must be designed so that its failure under excessive load would not impair the ability of the package to meet other requirements.

The location and construction of the lifting devices should be shown on the packaging drawings. Any other structural part of the package that could be used to lift the package must be rendered inoperable for lifting during transport or be designed with strength equivalent to that required for lifting attachments.

2.5.3.2 Tie-Down Devices

Review the design and evaluation of the tie-down devices that are a structural part of the package, their connection with the package body, and the package body in the local area around the tie-down devices. Verify that the design, testing, and analyses demonstrate that these devices comply with the requirements of 10 CFR 71.45(b):

- Any tie-down device which is a structural part of the package must be capable of withstanding, without generating stress in any material of the package in excess of its yield strength, a static force applied to the center of gravity of the package having a vertical component of 2 times the weight of the package with its contents, a horizontal component along the direction in which the vehicle travels of 10 times the weight of the package with its contents, and a horizontal component in the transverse direction of 5 times the weight of the package with its contents.

- A tie-down device which is a structural part of the package must be designed so that its failure under excessive load would not impair the ability of the package to meet other requirements.

The location and construction of the tie-down devices should be shown on the packaging drawings. Any other structural part of the package that could be used to tie down the package must be rendered inoperable for tying down the package during transport, or must be designed with strength equivalent to that required for tie-down devices.

2.5.4 General Considerations for Structural Evaluation of Packaging

The SAR should demonstrate that the analyses or tests used to evaluate the package under normal conditions of transport and hypothetical accident conditions have been adequately performed, including:

- The initial conditions (e.g., temperature, pressure, and residue heat) used are the most limiting for test or loading conditions of the packaging.

- The methods employed are appropriate for loading conditions considered and follow accepted practices and precepts.

- Interpretations of evaluation results are correct.

- The drop orientations considered in the evaluation are the most damaging. Note that the most damaging orientation for one component may not be the worst case for another component.

2.5.4.1 Evaluation by Analysis

If the structural evaluation is by analysis, the review should include the following:

- Verify that the SAR describes clearly the analysis models, methods, and results including all assumptions and input data used. The analysis model should adequately represent the geometry, boundary conditions, loading, material properties, and structural behavior of the packaging analyzed.

- Verify that the material model and properties are appropriate for the analyses. If the analysis is an elastic analysis, the material should also be modeled as an elastic material. If the analysis is inelastic, the actual material behavior or a conservative elastic-plastic material model representing the actual material should be used. The SAR should describe how the material properties were obtained and why the material model is appropriate for the loading conditions considered. For analysis involving large strains, the reviewer should verify that a stress-strain curve is used.

- Verify that the applied (force and displacement) boundary conditions in the analysis model are appropriate. For free-drop impact analyses, impact loads for package components are usually derived from a rigid body dynamic analysis of the package and used in a quasi-static analysis of the components. Verify that a dynamic amplification factor has been applied to the equivalent static load to account for all vibration effects that have been ignored in the rigid-body dynamic and quasi-static analysis. A summary of the quasi-static and dynamic analysis methods for impact analysis is provided in NUREG/CR-3966.

- Verify that the solution method is appropriate for the evaluation. If a computer program is used, the validity and reliability of the computer program should be verified. The SAR should describe the solution method, the bench marking results, and the quality assurance program for maintaining and using the computer code.

•   Verify that the most critical combinations of environmental and loading conditions are evaluated. At a minimum, the evaluation should cover all the initial and loading conditions listed in RG 7.8. In addition, verify that all critical free-drop orientations are evaluated in the SAR, assuming that the impact can be at any angle. In general, the drop orientations that should be evaluated consist of two groups: (1) drops that produce the highest g-loads to be used for impact analysis of the package components and (2) drops that attack the most vulnerable orientations and parts of the packaging (i.e., bolts, seals, valves and ports). The first group includes drops with the package center of gravity (c.g.) located directly above the center of the impact area. These drops are the end drops, the side drops, and the c.g.-over-corner drops. It also includes slap-down drops where the package c.g. is not directly above the impact area. A slap-down drop of a long package can produce a high g-load in the second impact due to a whipping action generated by the force of the first impact. The number of drops in the second group will depend on the vulnerable packaging components and their structural failure modes. Components vulnerable to impact loads should be protected from impacting directly by special design features such as recessed construction, protective cover plate, and impact limiter. The SAR should evaluate the consequences of all credible drops.

•   Verify that the analysis results are correctly interpreted or used to demonstrate adequate margins of safety of the structural design. The maximum stresses or strains should be compared to corresponding design-code allowables. Verify that the response of the package to loads and load combinations in terms of stress and strain to components and structural members is shown. The structural stability of individual members, as applicable, should be evaluated.

2.5.4.2 Evaluation by Test

If the structural evaluation is by test, the review should include at least the following:

•   Verify that the test procedures, test equipment and the impact pad are adequate for package impact testing. UCRL-ID-121673 provides guidelines for package drop testing including the use of reduced-scale models, which are commonly used for testing SNF packages.

•   Verify that the test specimen is fabricated using the same materials, methods, quality assurance, and inspection specifications as specified in the design. Any differences should be identified and the effects evaluated in the SAR. The specimens should include all safety components to be tested and components that are expected to have significant effects on the test results. Substitutes for the radioactive contents during the tests should have the same structural properties as the actual contents. The substitutes should have the same mass and same interaction with its surrounding packaging component as the actual contents. The same criteria should be used for all other simulated components to ensure that the simulated parts do not alter the test results. Verify that the scale model test specimen is properly scaled, fabricated, and instrumented (if applicable). In general, scale models do not provide reliable data to determine the leakage rate of the package. Verify that size effects of the scale model test article are not significant. The SAR should provide data to show that the size effect can be ignored, if a reduced scale model smaller than ¼ scale is used.

- Verify that the selected drop orientations are sufficient for a thorough test of all critical components of the package and the selection is supported by sound analysis or reasoning. The criteria in Section 2.5.4.1 for the selection of critical drop orientation for analysis can also be used here. The actual drop conditions and the resulting structural response or damage should be measured and recorded before, during, and after the tests. Verify that the methods and instruments are adequate for the measurements and the measurements are sufficient for describing the structural response or damage. Both interior and exterior damage of the test specimen should be included.

- Verify that all test results are evaluated and their structural integrity implication interpreted. The test conclusions should be valid and defensible. Unexpected or unexplainable test results indicating possible testing problems or previously unknown specimen behavior should be discussed and evaluated. In each test, the test measurements, damage, and observations should be consistent with each other. Inconsistencies should be identified and their possible causes explained in the SAR. Unreliable results should be identified and the need for additional tests assessed. If the package is permanently deformed or damaged, the possibility of further damage by subsequent test conditions should be evaluated. In addition, if the final damage is severe, the margin of safety of the package design against an unacceptable structural failure scenario such as a sudden or total collapse or rupture should be evaluated. If the final damage indicates the possibility of an imminent unstable structural failure, additional tests under the same test conditions should be performed to determine the repeatability of the result. If acceptance tests are performed on the specimen after the structural testing, the acceptance tests should be performed according to appropriate codes and standards.

2.5.5 Normal Conditions of Transport

The evaluation of the package performance under normal conditions of transport is based on the effects of the tests specified in 10 CFR 71.71. The ambient air temperature before and after the tests must remain near constant at that value between -29°C (-20°F) and +38°C (100°F) which are most unfavorable for the feature under consideration. The initial internal pressure within the containment system must be the MNOP unless a lower internal pressure consistent with the ambient temperature assumed to precede and follow the tests is more unfavorable. Separate specimens may be used for the free-drop test, the compression test, and the penetration test, if each specimen is subjected to the water spray test before being subjected to any of the other tests.

The SAR should show that the effectiveness of the package has not been reduced as a result of the normal conditions of transport, as specified by 10 CFR 71.43(f).

2.5.5.1 Heat

Verify that the heat loading condition will not compromise the structural integrity of the package.

Review the circumferential and axial deformations and stresses (if any) that result from differential thermal expansion. The evaluation should consider possible interferences resulting from a reduction in gap sizes. Verify that the stresses are within the limits for normal condition loads.

The evaluations should be based on the maximum ambient temperature and the MNOP in combination with the maximum internal heat load. For specified components of the package (e.g., elastomer seal, neutron shield material, etc.), review the maximum temperatures, and their effect on the operation of the package. Fatigue effects may be considered.

2.5.5.2 Cold

Verify that the evaluation for the cold condition is adequate. Confirm that the temperatures under the cold test condition are consistent with the thermal section.

The evaluations should consider the minimum internal pressure in combination with the minimum heat load, in combination with any residual fabrication stresses. Verify that differential thermal expansions that could result in possible geometric interfaces have been considered.

Verify that the stresses are within the limits for normal condition loads.

2.5.5.3 Reduced External Pressure

Determine that the SAR adequately evaluates the package design for the effects of reduced external pressure equal to 25 kPa (3.5 psi) absolute.

2.5.5.4 Increased External Pressure

Determine that the SAR adequately evaluates the package design for the effects of increased external pressure equal to 140 kPa (20 psi) absolute. Consider this loading condition in combination with minimum internal pressure. Consider the possibility of buckling (NUREG/CR-4554A).

2.5.5.5 Vibration

Determine that the SAR adequately evaluates the package design for the effects of vibration normally incident to transport. The SAR should provide a determination of the acceleration due to vibration by test or analysis. A fatigue analysis should be provided for highly stressed systems, considering the combined stresses due to vibration, temperature, and pressure loads. If closure bolts are reused, verify that the bolt preload is included in the fatigue evaluation (NUREG/CR-6007). Verify that a resonant vibration condition, which can cause rapid fatigue damage, is not present in any packaging component. The effect on package internals should be considered. References for vibration evaluation of transport packages include NUREG/CR-0128 and NUREG/CR-2146.

2.5.5.6 Water Spray

Review the package for the effects of the water spray test that simulates exposure to rainfall of approximately 5 cm (2 in) per hour for at least one hour. Verify that this test has no significant effects on material properties.

2.5.5.7 Free Drop

Review the package design for the effects of the free-drop test. Review procedures for impact are discussed in Section 2.5.4.

Review the evaluation of the closure lid bolt design, port cover plates, and other package components for the combined effects of free-drop impact force, internal pressures, thermal stress, and all other concurrently applied forces (e.g., seal compression force, bolt preload, etc.).

2.5.5.8 Corner Drop

This test applies only to fiberboard, wood, or fissile material rectangular packages not exceeding 50 kg (110 lb) and fiberboard, wood, or fissile material cylindrical packages not exceeding 100 kg (220 lb). This test is generally not applicable for SNF packages because their weight exceeds 100 kg (220 lb).

2.5.5.9 Compression

This test applies only to packages weighing up to 5000 kg (11,000 lb). This test is generally not applicable for SNF packages because their weight exceeds 5000 kg (11,000 lb).

2.5.5.10 Penetration

Review the evaluation of the package for the penetration condition. Verify that the most vulnerable orientations and locations of the package have been considered for this test condition.

2.5.6 Hypothetical Accident Conditions

The evaluation for hypothetical accident conditions must be based on sequential application of the tests specified in 10 CFR 71.73, in the order indicated, to determine their cumulative effect on a package. With respect to the initial conditions for the tests (except for the water immersion tests), the ambient air temperature before and after the tests must remain near constant at that value between -29°C (-20°F) and +38°C (100°F) which are most unfavorable for the feature under consideration. The initial internal pressure within the containment system must be the MNOP unless a lower internal pressure consistent with the ambient temperature assumed to precede and follow the tests is more unfavorable. Damage caused by the tests is cumulative, and the evaluation of the ability of a package to withstand any one test must consider the damage that resulted from the previous tests.

The package must have adequate structural integrity to satisfy the containment, shielding, subcriticality, and temperature requirements of 10 CFR Part 71. Generally, inelastic deformation of the containment closure system (e.g., bolts and flanges) is unacceptable for the containment evaluation. Deformation of other parts of the containment vessel may be acceptable if the containment boundary is not compromised. Deformation of shielding components is reviewed in the shielding evaluation. Deformation of components required for heat transfer and insulation is reviewed in terms of the thermal evaluation. Deformation of components required for subcriticality is reviewed in the criticality evaluations.

2.5.6.1 Free Drop

Review the evaluation of the package for the free-drop test. Verify that structural integrity has been evaluated for the drop orientation which produces the highest g-load and causes the most severe damage.

If a feature such as a tie-down component is a structural part of the package, it must be included in the drop-test configurations.

Review the evaluation of the closure lid bolt design, port cover plates, and other package components for the combined effects of free-drop impact force, internal pressures, thermal stress, and all other concurrently applied forces (e.g., seal compression, bolt preload, etc.).

Review the impact pad used for the free drop. Assure that an essentially unyielding pad, of adequate size, has been used. For a package with lead shielding, the effects of lead slump should be evaluated for the hypothetical accident condition free drop. The lead slump determined should be consistent with that assumed in shielding evaluation.

2.5.6.2 Crush

This test is only specified for packages with a mass not greater than 500 kg (1100 lb), density not greater than water, and radioactive contents greater than 1000 A_2, not as special form material.

This test is generally not applicable to SNF packages.

2.5.6.3 Puncture

Review the evaluation of the package for the puncture test. Verify that the orientation and location for which maximum damage would be expected have been considered. It should be noted that damages resulting from the drop test must be included when evaluating the puncture test.

Generally, thin-shelled packages are susceptible to puncture damage. Verify that puncture at oblique angles, near a support, at a valve, or at a penetration have been considered.

Although analytical methods are available for predicting punctures of plates, empirical formulas derived from puncture test results of laminated panels are usually used for determining the package surface layer thickness required for resisting punctures. The Nelm's formula developed specifically for package design provides the minimum thickness needed for preventing the puncture of the steel surface layer of a typical steel-lead-steel laminated cask wall. NUREG/CR-4554B provides an empirical formula for puncture evaluation based on empirical and analytical puncture studies. The formula is applicable for puncture at an angle normal to the surface and at a location away from a stiff support under the surface. The formula is conservative for solid packaging walls but may be nonconservative for punctures at an oblique angle, where the delivery of the puncture energy is more concentrated than in a right angle impact. Fortunately, there are few oblique punctures that can involve the total impact energy of a package. In general, oblique punctures may be critical for thin-shelled packages that require only a fraction of the total impact energy to penetrate the packaging wall.

2.5.6.4 Thermal

Verify that the package design is evaluated for a fully engulfing fire as specified in 10 CFR 71.73(c)(4). Any damage resulting from the free drop or puncture conditions must be incorporated into the initial condition of the package for the fire test. Confirm that the determination of the maximum pressure in the package during or after the test considers the temperatures resulting from the fire and any increase in gas inventory caused by thermal combustion or decomposition process. Verify that the maximum thermal stresses, which can occur either during or after the fire, are evaluated and are consistent with the Thermal Evaluation section of the SAR.

2.5.6.5 Immersion – Fissile Material

If water inleakage has not been assumed for the criticality analysis, review the evaluation of a damaged test specimen (i.e., after free drop, puncture, and fire) immersed under a head of water of at least 0.9 m (3 ft) in the attitude for which maximum leakage is expected.

2.5.6.6 Immersion – All Material

Review the evaluation of a separate, undamaged specimen subjected to water pressure equivalent to immersion under a head of water of at least 15 m (50 ft). For test purposes, an external pressure of water of 150 kPa (21.7 psi) gauge is considered to meet these conditions.

2.5.7 Special Requirement for Irradiated Nuclear Fuel Shipments

For a package of irradiated nuclear fuel with activity greater than 37 PBq (10^6 Ci), 10 CFR 71.61 requires that its undamaged containment system can withstand an external water pressure of 2 MPa (290 psi) for a period of not less than one hour without collapse, buckling, or inleakage of water. The SAR should provide analysis or test results to show that the containment structure will not collapse or buckle within one hour after the pressure is applied. This test applies only to the containment system. No structural support from other packaging components should be considered unless the component is an integral part of the containment system. The inleakage requirement has not been met if the stresses around the closure seal region exceed the yield stress limits.

2.5.8 Internal Pressure Test

For a package with a MNOP exceeding 35 kPa (5 psi) gauge, 10 CFR 71.85(b) requires that prior to first use the containment system be pressure tested at 150% of its MNOP. The analysis of this acceptance test should be provided in the SAR. The analysis should show that the package containment structure does not yield under the test pressure and the stresses are within the allowable stress limits set by the design code.

2.5.9 Appendix

The appendix may include a list of references, copies of any applicable references not generally available to the reviewer, computer code descriptions, input and output files, test results, and other appropriate supplemental information.

2.6 EVALUATION FINDINGS

The structural review should result in the following findings, as appropriate:

2.6.1 Description of Structural Design

The staff has reviewed the package structural design description and concludes that the contents of the application meet the requirements of 10 CFR 71.31

The staff has reviewed the codes and standards used in package design and find that they are acceptable.

2.6.2 Material Properties

To the maximum credible extent, there are no significant chemical, galvanic or other reactions among the packaging components, among package contents, or between the packaging components and the contents in dry or wet environment conditions. The effects of radiation on materials are considered and package containment is constructed from materials that meet the requirement of RGs 7.11 and 7.12.

2.6.3 Lifting and Tie-down Standards for All Packages

The staff has reviewed the lifting and tie-down systems for the package and concludes that they meet 10 CFR 71.45 standards.

2.6.4 General Considerations for Structural Evaluation of Packaging

The staff has reviewed the packaging structural evaluation and concludes that the application meets the requirements of 10 CFR 71.35.

2.6.5 Normal Conditions of Transport

The staff has reviewed the packaging structural performance under the normal conditions of transport and concludes that there will be no substantial reduction in the effectiveness of the packaging.

2.6.6 Hypothetical Accident Conditions

The staff has reviewed the packaging structural performance under the hypothetical accident conditions and concludes the packaging has adequate structural integrity to satisfy the subcriticality, containment, shielding, and temperature requirements of 10 CFR Part 71.

2.6.7 Special Requirement for Irradiated Nuclear Fuel Shipments

The staff has reviewed the containment structure and concludes that it will meet the 10 CFR 71.61 requirements for irradiated nuclear fuel shipments.

2.6.8 Internal Pressure Test

The staff has reviewed the containment structure and concludes that it will meet the 10 CFR 71.85(b) requirements for pressure test without yielding.

2.7 REFERENCES

ANSI N14.6   Institute for Nuclear Materials Management, ANSI N14.6, "Special Lifting Devices for Shipping Containers Weighing 10,000 Pounds (45000 kg) or More for Nuclear Materials," New York, NY, 1993.

B&PV Division 3 Code	American Society of Mechanical Engineers, "ASME Boiler and Pressure Vessel Code, Section III, Division 3, Containment Systems and Transport Packagings For Spent Nuclear Fuel and High Level Radioactive Waste," New York, NY, 1998.
NUREG-0612	U.S. Nuclear Regulatory Commission, "Control of Heavy Loads at Nuclear Power Plants," NUREG-0612, National Technical Information Service, Springfield, VA, July 1980.
NUREG/CR-0128	U.S. Nuclear Regulatory Commission, "Shock and Vibration Environments for a Large Shipping Container During Truck Transport (part II)," NUREG/CR-0128, U.S. Government Printing Office, Washington, D.C., August, 1978.
NUREG/CR-2146	U.S. Nuclear Regulatory Commission, "Dynamic Analysis to Establish Normal Shock and Vibration of Radioactive Material Shipping Packages, Volume 3: Final Summary Report," NUREG/CR-2146, Vol. 3, U.S. Government Printing Office, Washington, D.C., October 1983.
NUREG/CR-3854	U.S. Nuclear Regulatory Commission, "Fabrication Criteria for Shipping Containers," NUREG/CR-3854, Lawrence Livermore National Laboratory, Livermore, CA, March 1985.
NUREG/CR-3966	U.S. Nuclear Regulatory Commission, "Methods for Impact Analysis of Shipping Containers," NUREG/CR-3966, U.S. Government Printing Office, Washington, D.C., November 1987.
NUREG/CR-4554A	U.S. Nuclear Regulatory Commission, "SCANS (Shipping Cask Analysis System): A microcomputer Based Analysis System for Shipping Cask Design Review, Volume 6 – Theory Manual: Buckling of Circular Cylindrical Shells," NUREG/CR-4554, Vol. 6, U.S. Government Printing Office, Washington, D.C., February 1990.
NUREG/CR-4554B	U.S. Nuclear Regulatory Commission, "SCANS (Shipping Cask Analysis System): A microcomputer Based Analysis System for Shipping Cask Design Review, Volume 7 – Theory Manual: Puncture of Shipping Casks," NUREG/CR-4554, Vol. 7, U.S. Government Printing Office, Washington, D.C., February 1990.
NUREG/CR-6007	U.S. Nuclear Regulatory Commission, "Stress Analysis of Closure Bolts for Shipping Casks," NUREG/CR-6007, U.S. Government Printing Office, Washington, D.C., January 1993.

NUREG/CR-6322   U.S. Nuclear Regulatory Commission, "Buckling Analysis of Spent Fuel Basket," NUREG/CR-6322, U.S. Government Printing Office, Washington, D.C., May 1995.

RG 7.6   U.S. Nuclear Regulatory Commission, "Design Criteria for the Structural Analysis of Shipping Cask Containment Vessels," Regulatory Guide 7.6, Rev.1, U.S. Government Printing Office, Washington, D.C., March 1978.

RG 7.8   U.S. Nuclear Regulatory Commission, "Load Combinations for the Structural Analysis of Shipping Casks," Regulatory Guide 7.8, U.S. Government Printing Office, Washington, D.C., May 1977.

RG 7.11   U.S. Nuclear Regulatory Commission, "Fracture Toughness Criteria of Base Material for Ferritic Steel Shipping Cask Containment Vessels with a Maximum Wall Thickness of 4 Inches (0.1m)," Regulatory Guide 7.11, U.S. Government Printing Office, Washington, D.C., June 1991.

RG 7.12   U.S. Nuclear Regulatory Commission, "Fracture Toughness Criteria of Base Material for Ferritic Steel Shipping Cask Containment Vessels with a Wall Thickness Greater than 4 Inches (0.1m)," Regulatory Guide 7.12, U.S. Government Printing Office, Washington, D.C., June 1991.

UCRL-ID-121673   Mok, G.C., R.W. Carlson, S.C. Lu, and L.E. Fischer, "Guidelines for Conducting Impact Tests on Shipping Packages for Radioactive Material," UCRL-ID-121673, Lawrence Livermore National Laboratory, Livermore, CA, September 1995.

3 THERMAL REVIEW

3.1 REVIEW OBJECTIVE

The objective of this review is to verify that the thermal performance of the package has been adequately evaluated for the tests specified under normal conditions of transport and hypothetical accident conditions and that the package design satisfies the thermal requirements of 10 CFR Part 71.

3.2 AREAS OF REVIEW

The SAR must be reviewed for adequacy of the description and evaluation of the thermal design. Areas of review include the following:

3.2.1 Description of the Thermal Design
3.2.1.1 Packaging Design Features
3.2.1.2 Codes and Standards
3.2.1.3 Content Heat Load Specification
3.2.1.4 Summary Tables of Temperatures
3.2.1.5 Summary Tables of Pressures in the Containment Vessel

3.2.2 Material Properties and Component Specifications
3.2.2.1 Material Thermal Properties
3.2.2.2 Technical Specifications of Components
3.2.2.3 Thermal Design Limits of Package Materials and Components

3.2.3 General Considerations for Thermal Evaluations
3.2.3.1 Evaluation by Analyses
3.2.3.2 Evaluation by Tests
3.2.3.3 Confirmatory Analyses
3.2.3.4 Effects of Uncertainties

3.2.4 Evaluation of Accessible Surface Temperatures

3.2.5 Evaluation under Normal Conditions of Transport
3.2.5.1 Heat and Cold
3.2.5.2 Maximum Normal Operating Pressure
3.2.5.3 Maximum Thermal Stress

3.2.6 Evaluation under Hypothetical Accident Conditions
3.2.6.1 Initial Conditions
3.2.6.2 Fire Test
3.2.6.3 Maximum Temperatures and Pressures
3.2.6.4 Maximum Thermal Stresses

3.2.7 Appendix

3.3 REGULATORY REQUIREMENTS

Regulatory requirements of 10 CFR Part 71 applicable to the thermal review are as follows:

3.3.1 Description of the Thermal Design

The packaging must be described in sufficient detail to provide an adequate basis for its evaluation.
[10 CFR 71.31(a)(1), 10 CFR 71.33(a)(5), 10 CFR 71.33(a)(6), 10 CFR 71.33(b)(1), 10 CFR 71.33(b)(3), 10 CFR 71.33(b)(5), 10 CFR 71.33(b)(7), and 10 CFR 71.33(b)(8)]

The SAR must identify established codes and standards applicable to the thermal design.
[10 CFR 71.31(c)]

The thermal design must not depend on a mechanical cooling system to meet the containment requirements of 10 CFR 71.51(a). [10 CFR 71.51(c)]

3.3.2 Material Properties and Component Specifications

The package must be described in sufficient detail to provide an adequate basis for its evaluation.
[10 CFR 71.31(a)(1), 10 CFR 71.33(a)(5), and 10 CFR 71.33(b)(3)]

3.3.3 General Considerations for Thermal Evaluations

The package must be evaluated to demonstrate that it satisfies the thermal requirements specified in 10 CFR Part 71, Subpart E, under the conditions and tests of Subpart F. [10 CFR 71.31(a)(2), 10 CFR 71.35(a), and 10 CFR 71.41(a)]

3.3.4 Evaluation of Accessible Surface Temperatures

The package must be designed, constructed, and prepared for shipment so that the accessible surface temperature of a package in still air at 38°C (100°F) in the shade will not exceed 85°C (185°F) in an exclusive-use shipment. [10 CFR 71.43(g)] (Temperature limits for non-exclusive-use shipments are assumed not to apply to spent nuclear fuel (SNF) packages.)

3.3.5 Thermal Evaluation under Normal Conditions of Transport

The package design must be evaluated to determine the effects of the conditions and tests under normal conditions of transport. The ambient temperature preceding and following the tests must remain near constant at that value between -29°C (-20°F) and +38°C (100°F) which are the most unfavorable for the feature under consideration. The initial internal pressure within the containment system must be considered to be the maximum normal operating pressure (MNOP), unless a lower internal pressure consistent with the ambient temperature considered to precede and follow the tests is more unfavorable. The conditions and tests of 10 CFR 71.71(c)(1) and 10 CFR 71.71(c)(2) for heat and cold respectively are the primary thermal tests for normal conditions of transport. [10 CFR 71.71]

The package must be designed, constructed, and prepared for transport so that there will be no significant decrease in packaging effectiveness under the tests specified in 10 CFR 71.71 (normal conditions of transport). [10 CFR 71.43(f) and 10 CFR 71.51(a)(1)]

3.3.6 Thermal Evaluation under Hypothetical Accident Conditions

The package design must be evaluated to determine the effects of the conditions and tests under a hypothetical accident. This accident includes a sequence of incidents (impact, puncture, thermal, and immersion) on a package (the crush test is generally not applicable to packages for SNF). Except for the water immersion tests, the ambient temperature preceding and following the tests must remain constant at that value between -29°C (-20°F) and +38°C (100°F) which are the most unfavorable for the feature under consideration. The initial internal pressure within the containment system must be considered to be the MNOP, unless a lower internal pressure consistent with the ambient temperature considered to precede and follow the tests is more unfavorable. The 30-minute, 800°C (1475°F) fire test of 10 CFR 71.73(c)(4) on a damaged package is the primary thermal test for hypothetical accident conditions. [10 CFR 71.73]

3.4 ACCEPTANCE CRITERIA

3.4.1 Description of the Thermal Design

The regulatory requirements in Section 3.3.1 identify the acceptance criteria.

3.4.2 Material Properties and Component Specifications

In addition to the regulatory requirements identified in Section 3.3.2, the temperatures of the materials and components used in the package should not exceed their specified maximum allowable temperatures.

3.4.3 General Considerations for Thermal Evaluations

In addition to the regulatory requirements identified in Section 3.3.3, the models used in the thermal evaluation must be described in sufficient detail to permit an independent review, with confirmatory calculations, of the package thermal design.

3.4.4 Evaluation of Accessible Surface Temperature

The regulatory requirements in Section 3.3.4 identify the acceptance criteria.

3.4.5 Thermal Evaluation under Normal Conditions of Transport

The regulatory requirements in Section 3.3.5 identify the acceptance criteria.

3.4.6 Thermal Evaluation under Hypothetical Accident Conditions

The regulatory requirements in Section 3.3.6 identify the acceptance criteria.

3.5 REVIEW PROCEDURES

The following procedures are generally applicable to the thermal review of all SNF transportation packages. Since packages for shipment of SNF are generally intended to be shipped by exclusive-use, only exclusive-use shipments are assumed in the following SRP review procedures.

The thermal review is based in part on the descriptions and evaluations presented in the General Information and Structural Evaluation sections of the SAR. Similarly, results of the thermal review are considered in the review of the SAR sections on Structural Evaluation, Containment Evaluation, Shielding Evaluation, Criticality Evaluation, Operating Procedures, and Acceptance Tests and Maintenance Program. Examples of SAR information flow into, within, and from the thermal review are shown in Figure 3-1.

3.5.1 Description of the Thermal Design

3.5.1.1 Packaging Design Features

Review the general description of the package presented in the General Information section of the SAR and any additional description of the thermal design in the Thermal Evaluation section. Verify that the package description in the General Information section of the SAR includes:

• A description of any structural and mechanical means for the transfer and dissipation of heat

• The identity and volumes of receptacles containing coolant

• The MNOP of the containment system

• The maximum amount of content decay heat

• The identity and volumes of any coolants. Verify that the thermal design does not depend on the presence of a mechanical cooling system to ensure containment.

All text, drawings, figures, and tables describing the thermal features in the Thermal Evaluation section should be consistent with those of the General Information section as well as those used in the applicant's thermal evaluation. Particular emphasis should be placed on the consistency of the component dimensions, materials, and material properties.

3.5.1.2 Codes and Standards

Verify that the established codes and standards used in all aspects of the thermal design and evaluation of the package, including material properties and components, are identified.

3.5.1.3 Content Heat Load Specification

Verify that the maximum decay heat of the package contents reported in the Thermal Evaluation section of the SAR is consistent with that in the General Information section and that this heat load is appropriately considered in all thermal evaluations.

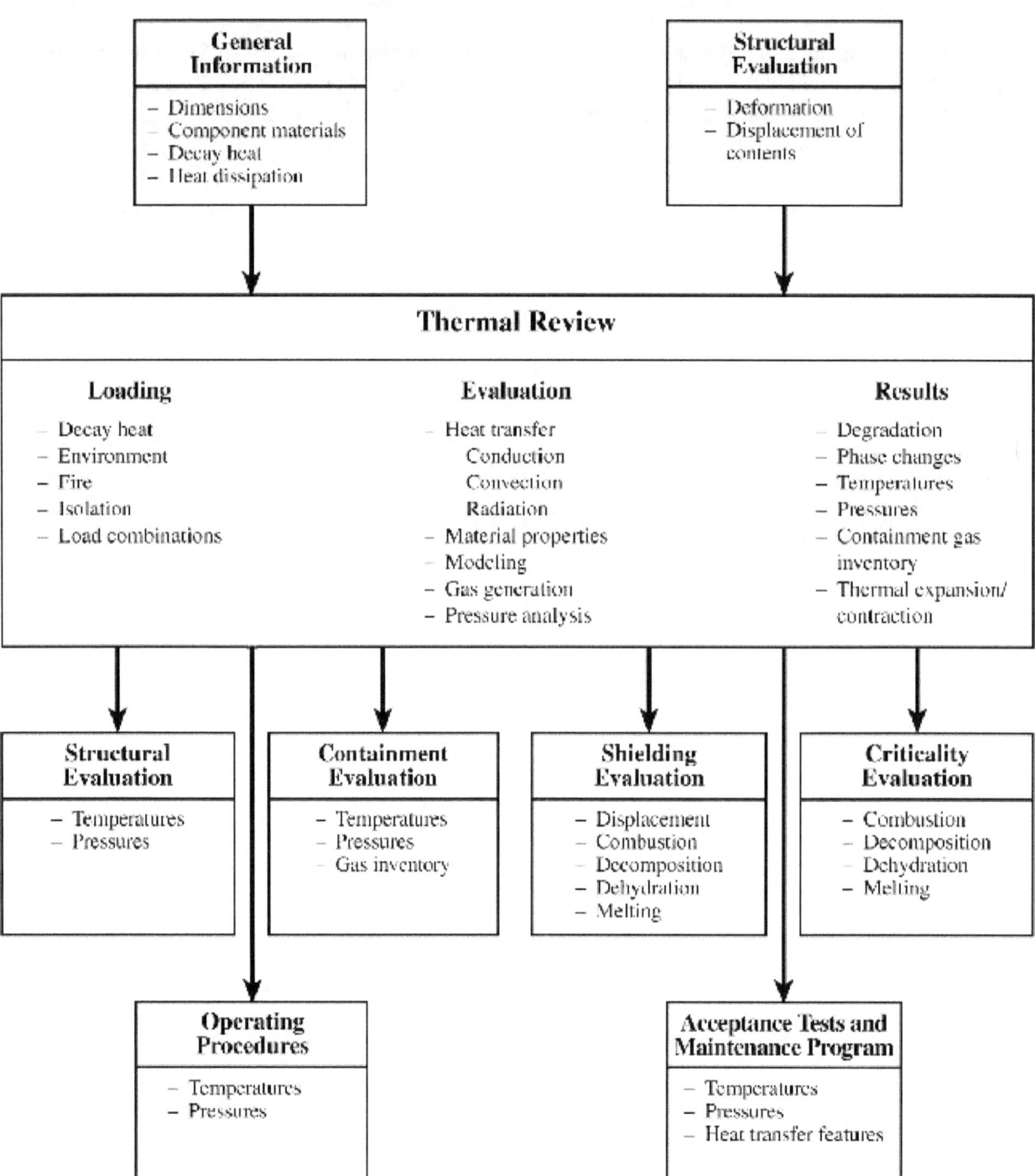

Figure 3-1 SAR Information Flow for the Thermal Review

Review the method in which the actual heat load is determined, and ensure that it is consistent with the SNF content specifications (e.g., burnup, enrichment, cooling time). If the heat load is based on the mass and decay energies of the contents, verify that it has been properly determined. The computer codes discussed in Section 5.5.2 for determination of neutron and gamma sources are often useful for calculating content decay heat loads. These codes are especially useful for SNF that contains a large number of radionuclide species.

3.5.1.4 Summary Tables of Temperatures

Confirm that summary tables of the temperatures of package components including, but not limited to, the fuel/cladding, basket, impact limiters, containment vessel, seals, shielding, and neutron absorbers are consistent with the temperatures presented in the General Information and Structural Evaluation sections of the SAR for the normal conditions of transport and hypothetical accident conditions. Confirm that the summary tables contain the design temperature limits for each of the components for the normal conditions of transport and hypothetical accident conditions. For the hypothetical accident condition fire, these summarized temperatures should additionally include the maximum temperatures after fire, the elapsed time from the beginning of the fire to the occurrence of these maximum temperatures, and the post-fire steady-state temperatures of each package component. Confirm that the temperatures and design temperature limit criteria for the package components are consistent throughout the appropriate sections of the SAR.

3.5.1.5 Summary Tables of Pressures in the Containment System

Verify that summary tables of the pressure in the containment system under the normal conditions of transport and hypothetical accident conditions are consistent with the pressures presented in the General Information, Structural Evaluation, Containment Evaluation, and Acceptance Tests and Maintenance Program sections of the SAR. The design pressure limits of the package components at the temperatures producing the pressures should be presented in the tables.

3.5.2 Material Properties and Component Specifications

3.5.2.1 Material Properties

Confirm that the thermal properties necessary to calculate thermal transport in the package as well as from the package to the environment are presented. These properties include, but are not limited to:

- thermal conductivity

- specific heat

- density

- thermal radiation emissivity of the package surfaces.

Verify that the thermal emissivities are appropriate for the specific package surface conditions and radiant energy spectrums for each thermal condition being evaluated. Confirm that the type of emittances (hemispherical vis-à-vis normal) are specified. The thermal radiation absorptivity on the external packaging surface may be conservatively assumed to be unity to compensate for changes in the package surface from dirt, weathering, and handling during its lifetime. Consideration of a proposed value of less than unity in the SAR should be based on the demonstration that controls and procedures will be in place to ensure such a value throughout the package lifetime. Periodic visual examination followed by paint touch-up or washing may be sufficient if the absorbtivity takes adequate account of weathering. These controls and procedures should appear in the Operating Procedures and Acceptance Tests and Maintenance Program sections of the SAR.

Verify that, for surrounding air and any fluids present within the package, the following additional properties are presented:

- viscosity

- Prandtl number.

Confirm that the given fluid properties are adequate for evaluating thermal convection parameters such as the Prandtl number (a dimensionless number defined as the ratio of the momentum diffusivity to the thermal diffusivity) which can be determined from the other thermal properties presented.

Confirm that the thermo-mechanical properties of any packaging material that may cause temperature-induced pressures and/or stresses within the package materials are presented. These properties include, but are not limited to:

- coefficient of thermal expansion

- modulus of elasticity

- Poisson's ratio.

The coefficient of thermal expansion is usually the linear coefficient for isotropic solids and the volumetric coefficient for fluids. For an isotropic material, the linear coefficient is one-third the volumetric coefficient.

Ensure that the structural properties that affect thermal stresses are consistent with the values reported in the Structural Evaluation section.

If a package material is anisotropic, confirm that the directional properties of, for example, the thermal conductivity, modulus of elasticity, and the linear expansion coefficient are provided.

Confirm that the temperatures at which phase changes, decomposition, dehydration and combustion will occur are presented, along with thermal and thermo-mechanical properties resulting from the change.

Confirm that the thermal properties used for the analyses of the package are appropriate for the material specified for the package in the General Information section and are consistent with those used in the Structural Evaluation section of the SAR. Verify that the sources of the thermal properties used in the SAR are referenced. Authoritative sources of material properties data include, but are not limited to, those that reference experimental measurements. In general, textbooks are an unacceptable source of material properties data. If the applicant experimentally measures the thermal properties of the material and components used in the package, ensure that the experiments are performed under an approved quality assurance program.

Confirm the appropriateness of the use of temperature-dependent thermal properties in an analysis of the package response to thermal loads. If the material properties are not presented as a function of temperature, verify that the value conservatively under- or over-predicts temperatures or stresses, as appropriate, compared to the equivalent temperature-dependent property.

3.5.2.2 Technical Specifications of Components

Verify that references for the technical specifications of package components such as O-rings, pressure relief valves, bolts, etc., are identified. Confirm that any temperature constraints on the function of the components are identified (such as the allowable stress in a bolt). Verify that the minimum allowable service temperature of all components is less than or equal to $-40°C$ ($-40°F$) unless a minimum heat load is specified (see Section 3.5.5.1).

3.5.2.3 Thermal Design Limits of Package Materials and Components

Confirm that the maximum allowable temperatures for each component that could affect the containment, shielding, and criticality functions of the package are specified.

Verify that the maximum allowable fuel/cladding temperature is justified. The justification should consider the fuel and clad materials, irradiation conditions (e.g., the absorbed dose, neutron spectrum, and fuel burnup), and the shipping environment including the fill gas. Other necessary considerations include the elapsed time from removal of the SNF from the core to its placement into the transportation packaging, its time duration in the packaging, and its post-transport disposition. Examples of temperature limits include, but are not limited to:

- the temperature limit for metal fuel should be less than the lowest melting point eutectic of the fuel

- the temperature limit on the irradiated clad in an inert gas environment as determined by creep, creep rupture, or diffusion controlled cavity growth (PNL-6189, UCID-21181), as appropriate.

Verify that the temperature range of the thermal and structural properties for each package material exceed the specified and predicted temperature limits for the material.

3.5.3 General Considerations for Thermal Evaluations

Thermal evaluations of the package can be performed by either analyses or tests, or by a combination of both. Because of their mass and cost and the difficulty of decay-heat simulation, SNF packages are normally evaluated by analysis. In addition, the use of analysis to evaluate the thermal performance of a package will allow the "margin of safety" in the package design to be determined.

Review the Structural Evaluation and Thermal Evaluation sections of the SAR to determine the response of the package to the normal conditions of transport and hypothetical accident conditions. Verify that the corresponding models used in the thermal analyses are consistent with these effects. For example, the package might have impact limiters or an external neutron shield that would be damaged during the structural and thermal tests of 10 CFR 71.73.

3.5.3.1 Evaluation by Analyses

Confirm that the methods of thermal analysis are identified and sufficiently described to permit a complete review and independent verification. The thermal analyses in the SAR can be based on simple calculations, spreadsheet-type analyses, or detailed computer simulations. The level of detail appropriate for each analysis, including assumptions, depends on many physical variables such as: the package materials, SNF decay heat, geometric complexity, and package component surface conditions. Ensure that each method of thermal analysis:

- is properly referenced or derived in the SAR as appropriate

- clearly and completely states the assumptions made in modeling heat sources and heat transfer paths and modes

- accurately represents the physical characteristics of the package consistent with the above discussed thermal design features (Section 3.5.1.1)

- uses appropriate thermal properties for the materials of construction (Section 3.5.2.1)

- uses appropriate expressions for conduction, convection, and thermal radiation among package components, and from the surfaces of the package to the environment

- correctly incorporates the appropriate specified temperature and thermal boundary conditions for the normal conditions of transport and hypothetical accident conditions.

For the 30-minute fire, the majority of the heat input to the package being tested will be radiative. For convective heat transfer, a convective heat transfer coefficient appropriate for the conditions that would exist if the package was exposed to the fire specified in 10 CFR 71.73, should be used. Flame velocities in an open pool fire may be used in determining the appropriate convective heat transfer coefficient. Flame velocities in open pool fires are discussed by Burgess and Fry (1990), Burgess (1987), and Schneider and Kent (1989). During the post-fire cooldown, natural convection should be assumed.

For the thermal analysis of steady-state normal conditions of transport (including the case for determining the accessible surface temperature), confirm the presence of either the Rayleigh number (product of the ratio of buoyancy force to the viscous force and the ratio of the momentum diffusivity to the thermal diffusivity) for the air at the surfaces of the package, or information sufficient to calculate the Rayleigh number. For the 30-minute fire, the appropriate radiative and convective heat transfer coefficients should be supported by the necessary correlations. Any correlations used in the analysis should be properly explained and justified.

Confirm that the assumptions about contact resistance at material interfaces, energy transport across gaps or enclosures, etc. are presented and appropriate. For example, the assumption of a maximum contact resistance between component surfaces during steady-state conditions or during a post-fire cooldown and the assumption of no contact resistance between component surfaces during a fire will result in maximizing the calculated component temperatures for normal conditions of transport and hypothetical accident conditions.

Under the conditions where any of the cask component temperatures are close (within 5%) to their limiting values during an accident or the MNOP is within 10% of its design basis pressure, or any other special conditions, the applicant should consider, by analysis, the potential impact of the fission gas in the canister to the cask component temperature limits and the cask internal pressurization.

In the case of computer analysis, the applicant may use standard off-the-shelf software or develop a computer code to perform a specific analysis. Verify that the code has been benchmarked and is maintained and operated under a quality assurance program. Verify that the code has been appropriately used. Ensure that the SAR includes appropriate code input and output files to enable a detailed review of the analysis.

3.5.3.2 Evaluation by Tests

For those results determined by tests, verify that a description of the test package, the test facility, and the test procedures used for simulating either the normal conditions of transport or hypothetical accident conditions are reported in adequate detail. Confirm that the test package was fabricated, the test facility operated, and the test results evaluated under proper quality assurance programs.

Review the ability of both the test facilities and test procedures to meet the range of specified temperatures: from -29°C (-20°F) to 38°C (100°F) for normal conditions of transport and both 38°C (100°F) and 800°C (1475°F) for hypothetical accident conditions. Confirm that the facilities can simulate the specified heat-transfer boundary conditions:

- incident heat fluxes equivalent to or exceeding the specified insolation requirements during the normal conditions of transport or the post-fire environment for hypothetical accident conditions

- incident heat fluxes equivalent to or exceeding the specified convective and radiative heat transfer environment, including specified emissivities, for a minimum 30-minute period representing the hypothetical accident condition fire

- an environment that assures an adequate supply and circulation of oxygen for initiating and naturally terminating the combustion of any burnable package component.

Confirm that the test package, with a simulated package contents and any attached test instrumentation or hardware, adequately simulates the thermal behavior of the actual package design.

Verify that the locations of the temperature and heat flux sensing devices are shown on figures in the SAR. Verify that the temperature sensing devices are placed on the test package:

- on applicable components

- in such a manner that they do not unduly affect local temperatures

- in locations where maximum temperatures are expected and where other temperatures need to be determined

- in locations that permit reasonable interpolation or extrapolation of measured temperatures for estimating temperatures in unmonitored regions of the package.

The applicable components include, but are not limited to, the containment vessel, fuel basket, seals, radiation shielding, criticality controls, and impact limiters. Confirm that the temperature sensing devices are measuring the temperature of the component, not that of the component environment.

Verify that the test time is sufficient for temperatures to reach steady-state conditions under normal conditions of transport or their peak following cessation of the hypothetical accident condition fire. To the extent that specified boundary conditions, the decay heat of the contents, or specified temperatures are not achieved during a test, verify that the evaluations include appropriate corrections to the temperature data.

Additional guidelines on reviewing thermal tests under hypothetical accident conditions are presented in NUREG/CR-5636, SAND85-0196, Hovingh and Carlson (1994), and UCRL-ID-110445.

3.5.3.3 Confirmatory Analyses

The rigor required of the confirmatory analysis will depend on the size of the margin between the maximum package component temperatures determined by the applicant and the maximum temperature limit specified for a material or component or the regulatory limit determined by the type of shipment. A conservative method of analysis of the fire portion of the hypothetical accident is to mathematically apply an 800°C (1475°F) surface temperature for 30 minutes to the package with the appropriate initial temperature distribution and content decay heat. This will eliminate the questions about the flame velocity and its effect on the convection heat input into the package. The analysis will still require the appropriate boundary conditions during cooldown to calculate the maximum component temperatures.

3.5.3.4 Effects of Uncertainties

Verify that the thermal evaluations appropriately address the effects of uncertainties in thermal and structural properties of materials, test conditions and diagnostics, and analytical methods, as applicable.

3.5.4 Evaluation of Accessible Surface Temperatures

Determine that the thermal model used for the calculation of the accessible surface temperature is presented in the SAR. This model should consist of a heat balance at the surface of the package in which the decay heat from the contents at the surface of the package is equal to the convective and radiative heat loses to the environment at an ambient temperature of 38°C (100°F).

If the maximum surface temperature of a package exceeds the regulatory limit, a personnel barrier can be placed around the package. This personnel barrier becomes the accessible package surface. The thermal impedance of the barrier should be considered when determining the package temperatures for normal conditions of transport.

Confirm that the maximum accessible surface temperature determined by the applicant is consistent with the General Information section of the SAR.

When appropriate, perform an independent analysis as described in Section 3.5.3.6 to confirm the maximum accessible surface temperature determined by the applicant.

Ensure that the maximum temperature of the accessible package surface does not exceed 85°C (185°F) for exclusive use shipment when the package is subjected to the heat conditions of 10 CFR 71.43(g).

3.5.5 Thermal Evaluation under Normal Conditions of Transport

3.5.5.1 Heat and Cold

Confirm that the thermal evaluation demonstrates that the tests for normal conditions of transport do not result in significant reduction in packaging effectiveness, including:

- degradation of the heat-transfer capability of the packaging (such as creation of new gaps between components)*

- changes in material conditions or properties (e.g., expansion, contraction, gas generation, and thermal stresses) that affect the structural performance

- changes in the packaging or contents that affect containment, shielding, or criticality such as thermal decomposition or melting of materials

- ability of the package to withstand the tests under hypothetical accident conditions.

Verify that the SAR properly determines the maximum temperatures of the package components during normal conditions of transport when the package is in 38°C (100°F) still air with insolation, according to the table in 10 CFR 71.71(c)(1), and the content heat load is the maximum allowable. Temperatures of special interest include, but are not limited to, those of the fuel/cladding, containment vessel, seals, shielding, criticality controls, and impact limiters. Confirm that the volume-averaged temperature of gases are determined. Verify that the results are consistent with the General Information and Structural Evaluation sections of the SAR.

Ensure that the SAR determines the minimum temperatures of the package components during normal conditions of transport when the package is in -40°C (-40°F) still air without insolation and the content heat load is the minimum allowable. If the SAR does not restrict the minimum heat load, the package should be considered at a uniform temperature of -40°C (-40°F). Verify that these temperatures are consistent with the Structural Evaluation section of the SAR.

Confirm that the maximum and minimum temperatures do not exceed their allowable limits, as specified in Section 3.5.2.3.

3.5.5.2 Maximum Normal Operating Pressure (MNOP)

Confirm that the SAR determines the maximum normal operating pressure when the package has been subjected to the heat condition for one year, as specified in 10 CFR 71.71(c)(1). Ensure that the evaluation has considered all possible sources of gases, such as:

- gases present in the package at closure

- fill gas released from the SNF rods

- fission product gases released from the SNF

- saturated vapor from material in the containment vessel including water vapor desorbed from the containment system components or the package contents

- helium from the a-decay of the SNF contents

- hydrogen and other gases from radiolysis or chemical reactions (e.g., sodium-water)

- hydrogen and other gases from the dehydration, combustion, or decomposition of package components.

Guidance on release of fill gas and fission product gas for PWR and BWR fuel is provided in Table 4-1.

Verify that MNOP is consistent with the Structural Evaluation section of the SAR.

If the package has any confined volumes other than the containment vessel (e.g., coolant tanks), confirm that their pressures are properly determined and consistent with the Structural Evaluation.

3.5.5.3 Maximum Thermal Stresses

There are two sources of thermal stresses. These stresses can be caused by either spatial temperature gradients in constrained package components or by interference between components due to different thermal expansions of the components.

Two cases should be investigated for interferences between package components assembled at "room" temperature. These include:

- a steady-cold (-40°C [-40°F]) environment with maximum SNF decay heat without insolation

- a steady-hot (38°C [100°F]) environment with maximum SNF decay heat with insolation

Confirm that the dimensions of the package components, and the clearances or interferences, for the above cases are presented in the SAR. Verify that an appropriate method for estimation of the stresses from the interferences between components is described. Verify that the stresses from any interferences between components are consistent with the Structural Evaluation section of the SAR.

3.5.6 Thermal Evaluation under Hypothetical Accident Conditions

Verify that the package has been evaluated to demonstrate the effects of the tests for hypothetical accident conditions.

3.5.6.1 Initial Conditions

Prior to the fire test, the package must be evaluated for the effects of the drop, crush (if applicable), and puncture tests. Ensure that the initial physical condition of the package represented in the thermal evaluations under hypothetical accident conditions is consistent with these results from the Structural Evaluation section of the SAR.

Verify that the SAR justifies the most unfavorable initial ambient temperature between -29°C (-20°F) and +38°C (100°F). Unless the package is susceptible to increased structural damage at lower temperatures, the initial ambient temperature should be 38°C (100°F). Verify that the initial steady-state temperature distribution is consistent with the results from the thermal evaluations under normal conditions of transport.

Confirm that the initial internal pressure of the package is the MNOP unless a lower internal pressure, consistent with the initial ambient temperature, is more unfavorable. Similarly, confirm that the internal heat load of the SNF contents is at its maximum allowable power unless a lower power, consistent with the temperature and pressure, is more unfavorable.

3.5.6.2 Fire Test

Verify that the package is exposed to the 800°C (1475°F) fire environment for a minimum of 30 minutes and that surface and fire emissivity are greater than or equal to 0.8 and 0.9, respectively. Confirm that the flame velocities are specified and appropriate for the hydrocarbon fire and that the appropriate correlation for convection in the fire is used as a boundary condition.

Verify that after the fire:

- No artificial cooling is applied to the package

- The package is subjected to full insolation

- The evaluation continues until the post-fire, steady-state condition is achieved

- An adequate supply of oxygen is continued throughout this period

- All combustion is allowed to proceed until it terminates naturally.

3.5.6.3 Maximum Temperatures and Pressures

Verify that the SAR appropriately evaluates the transient peak temperatures of the package components as a function of time after the fire. The maximum temperatures in the components will occur following cessation of the fire, with the delay time increasing with the distance inward from the package surface. Verify also that the SAR determines the maximum temperatures of the post-fire, steady-state condition.

Confirm that the maximum temperatures do not exceed the maximum allowable temperature limits.

Verify that the evaluation of the maximum pressure in the containment vessel is based on MNOP (Section 3.5.5.2) as it is affected by the fire-caused increases in package component temperatures. Confirm that possible increases in gas inventory (e.g., from fuel rod failure) have been considered in the pressure determination.

If the package has any confined volumes other than the containment vessel (e.g.., coolant tanks), confirm that their pressures are properly determined.

Verify that maximum temperatures and pressures are consistent with the Structural and Containment Evaluations.

3.5.6.4 Maximum Thermal Stresses

Verify that the SAR evaluates the thermal stresses. The maximum interference between components in a package during a hypothetical thermal accident usually occur during the post-fire cooldown. Where the components are concentric, the tensile stresses occur in the outer component while the stresses in the inner components are usually compressive.

Ensure that the maximum thermal stresses are consistent with those in the Structural Evaluation section of the SAR.

3.5.7 Appendix

The appendix may include a list of references, copies of any applicable references not generally available to the reviewer, computer code descriptions, input and output files, test facility and instrumentation descriptions, test results, special analyses, and other appropriate supplemental information.

3.5.7.1 Justification for Assumptions or Analytical Procedures

Confirm that the applicant has stated and justified all assumptions used in the evaluation of the package. Review the appropriateness of and justification for the applicant's assumptions and analytical procedures.

3.5.7.2 Computer Program Description

Confirm that the applicant describes all the computer programs used in the thermal evaluation of the package. Verify that the space dimensionality and method of analysis (finite difference, finite element, etc.) are identified. Verify that the range of applications and phenomena (linear, nonlinear; steady state, transient; etc.) as well as the material properties and material models (isotropic, anisotropic, etc.) are described. Verify that the various types of initial boundary conditions and thermal loads are described. Verify that solution techniques (direct or iterative for steady state; explicit, implicit, etc. for transient) are identified. Also verify that any other capabilities (enclosure radiation with view factor calculation, thermal stress analysis, etc.) that are applicable to the applicant's thermal evaluation are identified and described. Verify that the computer programs are appropriate for the problem to which they are applied by the applicant.

3.5.7.3 Computer Input and Output Files

Confirm that the applicant has submitted annotated input files, as applicable, for each problem (maximum accessible surface temperature, normal conditions of transport, calculation of initial temperature distribution for hypothetical accident, initial temperature distribution for analysis of thermal hypothetical accident) analyzed using a computer code. Confirm that the applicant has submitted annotated output files, as applicable, for each problem (maximum accessible surface temperature, normal conditions of transport, calculation of initial temperature distribution for hypothetical accident conditions, and temperature distribution histories for the thermal hypothetical accident condition during and following the 30-minute fire, until all the package component temperatures have reached their maxima).

3.5.7.4 Description of Test Facilities

Verify that the facilities used for performing thermal tests are described. The description shall include, but is not limited to:

- the type of facility (furnace, pool fire, etc.)

- the method of heating the package (gas burners, electrical heaters, etc.).

The description of a furnace facility should include the volume and emissivity of the furnace interior as well as the method of measuring the interior temperature. The oxygen concentration in a furnace test should be consistent with that of a hydrocarbon-fuel fire.

For a pool fire facility, the size of the fire relative to the size of the package shall be specified. Verify that the fire dimension conforms to the regulatory requirement that the fire thickness extend horizontally at least one meter (but not more than three meters) beyond any external surface of the package. The package will be positioned one meter above the surface of the fuel source. Verify that the method of support of the package in a test facility is described and an analysis of the heat loss from the package through the support to "ground" is presented. Review that the analysis of the heat loss from the package through the support is appropriate.

Confirm that the sensors used to measure heat flux and temperature are identified and described. Verify that the applicable operating ranges of the sensors are presented. Verify that the perturbation by the sensor (due to heat losses along thermocouple leads, shadowing by heat flux measuring devices, etc.) on the quantity to be measured (temperature, heat flux, etc.) is presented and quantified. Review that the heat flux and temperature sensors are appropriate and that the measurements are corrected for the perturbations by the sensors on the quantity to be measured. Verify that if calorimeters are used to measure heat flux, the calorimeter readings are corrected to account for the difference in thermal inertia between the calorimeter and the package. Verify that the method of correction of the calorimeter reading is presented and review the method for appropriateness.

3.5.7.5 Test Results

Verify that test measurements including temperatures (or temperature histories) and flux (or flux histories) are presented. Verify that the corrected test results are presented and that appropriate methods are used to obtain these corrections. Verify that, for the thermal portion of the hypothetical accident, the time at which the 30-minute test starts and ends is clearly noted. Verify that the measurements (and corrected results) are continued until steady state occurs (for tests for normal conditions of transport) or until the maximum temperature occurs in all the package components (for test of the thermal portion of the regulatory hypothetical accident).

Verify that photographs of the package components prior to and following the tests are presented. Verify that photographs of regions of components with thermal damage (such as charring of the insulation, damage to O-rings, etc.) are presented.

3.5.7.6 Applicable Supporting Documents or Specifications

Verify that the applicable sections from reference documents are included. These documents may include the test plans used for the thermal tests, the thermal specifications of O-rings and other components, and the documentation of the thermal properties of non-ASME approved materials used in the package.

3.5.7.7 Special Analyses

Frequently, thermally driven special processes will occur in a package. These processes may include, but are not limited to:

• generation of gases within the containment system

- effects of phase changes on package materials

- combustion, decomposition or dehydration of package materials.

The production of gases (e.g., hydrogen by radiolysis) or thermal decomposition of materials (e.g., a neutron shield) may occur in the package. Phase changes of material resulting in a decrease of the material density occurring in the containment system or in a lead shield can result in a pressure increase in the system. The tests under hypothetical accident conditions may cause combustion, decomposition or dehydration of components such as an impact limiter or the neutron shield material.

Confirm that the applicant has identified all thermally driven special processes that will occur in the package. Verify that the applicant has stated and justified all assumptions used in the quantification and evaluation of these special processes. Review the appropriateness of and justification for the applicant's assumptions and analytical procedures. Verify that the results are incorporated in the appropriate subsections of the Thermal Evaluation section.

Other supplemental calculations may be required to support evaluations presented in the Thermal Evaluation section. Verify that all such special analyses meet the goals discussed in Section 3.5.3.1.

3.6 EVALUATION FINDINGS

The thermal review should result in the following findings, as appropriate:

3.6.1 Description of the Thermal Design

The staff has reviewed the package description and evaluation and concludes that they satisfy the thermal requirements of 10 CFR Part 71.

3.6.2 Material Properties and Component Specifications

The staff has reviewed the material properties and component specifications used in the thermal evaluation and concludes that they are sufficient to provide a basis for evaluation of the package against the thermal requirements of 10 CFR Part 71.

3.6.3 General Considerations for Thermal Evaluations

The staff has reviewed the methods used in the thermal evaluation and concludes that they are described in sufficient detail to permit an independent review, with confirmatory calculations, of the package thermal design.

3.6.4 Evaluation of Accessible Surface Temperature

The staff has reviewed the accessible surface temperatures of the package as it will be prepared for shipment and concludes that they satisfy 10 CFR 71.43(g) for packages transported by exclusive-use vehicle.

3.6.5 Evaluation under Normal Conditions of Transport

The staff has reviewed the package design, construction, and preparations for shipment and concludes that the package material and component temperatures will not extend beyond the specified allowable limits during normal conditions of transport consistent with the tests specified in 10 CFR 71.71.

3.6.6 Evaluation under Hypothetical Accident Conditions

The staff has reviewed the package design, construction, and preparations for shipment and concludes that the package material and component temperatures will not exceed the specified allowable short time limits during hypothetical accident conditions consistent with the tests specified in 10 CFR 71.73.

3.7 REFERENCES

Burgess 1987	Burgess, M.H., "Heat Transfer Boundary Conditions in Pool Fires," *Packaging and Transportation of Radioactive Material (PATRAM)*, Vol. II., STI/PUB/718, IAEA, Vienna, 1987, pp. 423-431.
Burgess and Fry 1990	Burgess, M.H., and C.J. Fry, "Fire Testing for Package Approval," *Int. J. Radioactive Materials Transport*, Vol. 1, No. 1, Nuclear Technology Publishing, Ashford, Kent, England, 1990, pp. 7-16.
60 FR 50247	U.S. Nuclear Regulatory Commission, "Compatibility With the International Atomic Energy Agency (IAEA)," *Federal Register*, FR 50247, U.S. Government Printing Office, Washington, D.C., September 28, 1995.
SAND85-0196	Gregory, J.J., R. Mata, and N.R. Keltner, "Thermal Measurements in a Series of Large Pool Fires," SAND85-0196, TTC-0659, UC-71, Sandia National Laboratories, Albuquerque, NM, August 1987.
Hovingh and Carlson 1994	Hovingh, J., and R.W. Carlson, "Thermal Testing Transport Packages for Radioactive Materials - Reality vs. Regulation," ASME 1994 Pressure Vessel & Piping Conference, Minneapolis, MN, June 1994.
NUREG/CR-5636	U.S. Nuclear Regulatory Commission, "Fire and Furnace Testing of Transportation Packages for Radioactive Materials," NUREG/CR-5636, U.S. Government Printing Office, Washington, D.C., January 1999.
PNL-6189	Levy, I.S., et al., "Recommended Temperature Limits for Dry Storage of Spent Light Water Reactor Zircaloy-Clad Fuel Rods in Inert Gas," PNL-6189, Pacific Northwest Laboratory, Richland, WA, May 1987.
Schneider and Kent	Schneider, M.E., and L.A. Kent, "Measurements of Gas Velocities and

1989	Temperatures in a Large Open Pool Fire," *Fire Technology*, Vol. 25, February 1989, pp. 51-80.
UCID-21181	Schwartz, M.W. and M.C. Witte, "Spent Fuel Cladding Integrity During Dry Storage," UCID-21181, Lawrence Livermore National Laboratory, Livermore, CA, September 1987.
UCRL-ID-110445	VanSant, J.H., R.W. Carlson, L.E. Fischer, and J. Hovingh, "A Guide for Thermal Testing Transport Packages for Radioactive Material - Hypothetical Accident Conditions," UCRL-ID-110445, Lawrence Livermore National Laboratory, Livermore, CA, February 9, 1993.

4 CONTAINMENT REVIEW

4.1 REVIEW OBJECTIVE

The objective of this review is to verify that the package design satisfies the containment requirements of 10 CFR Part 71 under normal conditions of transport and hypothetical accident conditions.

4.2 AREAS OF REVIEW

The SAR should be reviewed for adequacy of the description and evaluation of the containment design. Areas of review include the following:

4.2.1 Description of Containment System
4.2.1.1 Containment Boundary
4.2.1.2 Codes and Standards
4.2.1.3 Special Requirements for Damaged Spent Nuclear Fuel

4.2.2 Containment under Normal Conditions of Transport
4.2.2.1 Pressurization of Containment Vessel
4.2.2.2 Containment Criteria
4.2.2.3 Compliance with Containment Criteria

4.2.3 Containment under Hypothetical Accident Conditions
4.2.3.1 Pressurization of Containment Vessel
4.2.3.2 Containment Criteria
4.2.3.3 Compliance with Containment Criteria

4.2.4 Appendix

4.3 REGULATORY REQUIREMENTS

Regulatory requirements of 10 CFR Part 71 applicable to the containment review are as follows:

4.3.1 Description of Containment System

The packaging must be described in sufficient detail to provide an adequate basis for its evaluation. [10 CFR 71.31(a)(1), 10 CFR 71.33(a)(4), 10 CFR 71.33(a)(5), 10 CFR 71.33(b)(1), 10 CFR 71.33(b)(3), 10 CFR 71.33(b)(5), and 10 CFR 71.33(b)(7)]

The SAR must identify established codes and standards applicable to the containment design. [10 CFR 71.31(c)]

The package must include a containment system securely closed by a positive fastening device that cannot be opened unintentionally or by a pressure that may arise within the package. [10 CFR 71.43(c)]

The package must be made of materials and construction that assure that there will be no significant chemical, galvanic, or other reaction. [10 CFR 71.43(d)]

Any valve or similar device on the package must be protected against unauthorized operation and, except for a pressure relief valve, must be provided with an enclosure to retain any leakage. [10 CFR 71.43(e)]

Spent fuel, with plutonium in excess of 0.74 TBq (20 Ci) per package, in the form of debris, particles, loose pellets, or fragmented rods or assemblies must be packaged in a separate inner container (second containment system) in accordance with 10 CFR 71.63(b). [10 CFR 71.63]

4.3.2 Containment under Normal Conditions of Transport

The package must be evaluated to demonstrate that it satisfies the containment requirements of 10 CFR Part 71, Subpart E, under the conditions and tests of Subpart F. [10 CFR 71.31(a)(2), 10 CFR 71.35(a), and 10 CFR 71.41(a)]

A package must meet the containment requirements of 10 CFR 71.43(f) and 10 CFR 71.51(a)(1) under the tests specified in 10 CFR 71.71 (normal conditions of transport), with no dependence on filters or a mechanical cooling system. [10 CFR 71.51(c)]

The package may not incorporate a feature intended to allow continuous venting during transport. [10 CFR 71.43(h)]

4.3.3 Containment under Hypothetical Accident Conditions

The package must be evaluated to demonstrate that it satisfies the containment requirements of 10 CFR Part 71, Subpart E, under the conditions and tests of Subpart F. [10 CFR 71.31(a)(2), 10 CFR 71.35(a), and 10 CFR 71.41(a)]

A package must meet the containment requirements of 10 CFR 71.51(a)(2) for hypothetical accident conditions, with no dependence on filters or a mechanical cooling system. [10 CFR 71.51(c)]

4.4 ACCEPTANCE CRITERIA

4.4.1 Description of Containment System

In addition to the regulatory requirements identified in Section 4.3.1, the containment system should be designed and constructed in accordance with Section III, Division 3, ASME Boiler and Pressure Vessel (B&PV Division 3) Code. Alternatives codes should be justified in the SAR.

In addition to the regulatory requirements identified in Section 4.3.1, leakage from the containment system should be determined in accordance with ANSI N14.5.

The codes, standards, and criteria for the inner containment system should generally be the same as those of the outer containment system. Justification for differences should be presented in the SAR.

In addition to the regulatory requirements identified in Section 4.3.1, damaged fuel should be canned to facilitate handling and to confine gross fuel particles to a known subcritical volume under normal conditions of transport and hypothetical accident conditions.

4.4.2 Containment under Normal Conditions of Transport

In addition to the regulatory requirements identified in Section 4.3.2, combustible gases should not exceed 5% of the free gas volume in any confined region of the package while the containment vessel is sealed and under normal transport conditions. The SAR should identify the allowable normal conditions of transport volumetric leakage rates in accordance with ANSI N14.5.

4.4.3 Containment under Hypothetical Accident Conditions

In addition to the regulatory requirements identified in Section 4.3.3, combustible gases should not exceed 5% of the free gas volume in any confined region of the package while the containment vessel is sealed and under hypothetical accident conditions. The SAR should identify the allowable hypothetical accident conditions volumetric leakage rates in accordance with ANSI N14.5.

4.5 REVIEW PROCEDURES

The following procedures are generally applicable to the containment review of all SNF transportation packages.

The containment review is based in part on the descriptions and evaluations presented in the General Information, Structural Evaluation, and Thermal Evaluation sections of the SAR and follows the sequence established to evaluate the packaging against applicable 10 CFR Part 71 requirements. Similarly, results of the containment review are considered in the review of the SAR sections on Operating Procedures and Acceptance Tests and Maintenance Program. Examples of SAR information flow into, within, and from the containment review are shown in Figure 4-1.

4.5.1 Description of the Containment System

4.5.1.1 Containment Boundary

Review the General Information section of the SAR and any additional description of the containment system presented in the Containment Evaluation section. All drawings, figures, and tables that describe containment features should be consistent with the evaluation.

Verify that the SAR provides a complete description of the containment boundary, including, as applicable, the containment vessel, welds, seals, lids, cover plates, valves, and other closure devices. The containment boundary should be clearly depicted in a figure or sketch. Ensure that all components of the containment boundary are shown in the drawings.

Confirm that the following information regarding components of the containment boundary is consistent with that presented in the Structural Evaluation and Thermal Evaluation sections of the SAR:

- Materials of construction

- Welds

- Applicable codes and standards (e.g., ASME B&PV Division 3 Code specifications for the vessel)
  Figure 4-1 SAR Information Flow for the Containment Review.

- Bolt torque required to maintain positive closure

- Maximum allowable temperatures of components, including seals

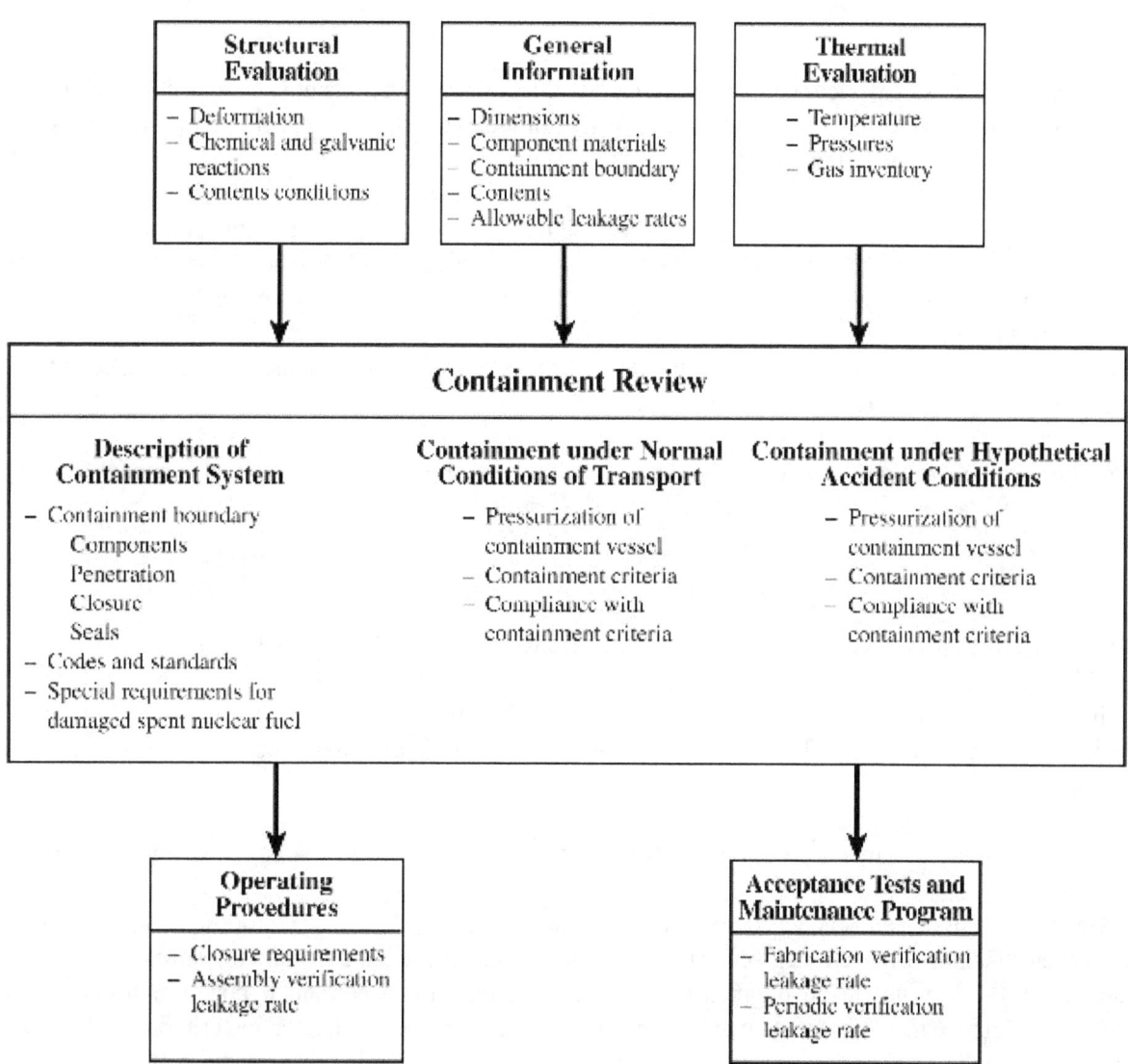

- Temperatures of components under normal conditions of transport and hypothetical accident conditions.

Verify that all containment boundary penetrations and their method of closure are described in detail. Performance specifications for components such as valves and O-rings should be documented, and no device may allow continuous venting. Any valve or similar device on the package must be protected against unauthorized operation and must be provided with an enclosure to retain any leakage. Cover plates and lids should be recessed or otherwise protected. Compliance with the permitted release limit may not depend on filters.

Confirm that all closure devices can be leak tested. If fill, drain, or test ports utilize quick connect valves, ensure that such valves do not preclude leakage testing of their cover-plate seals, providing such seals form part of the containment boundary.

Determine that the seal material is compatible for its intended use and that no galvanic or chemical reaction will occur between the seal and the packaging or its contents (NRC Bulletin 96-04). If penetrations are closed with two seals (e.g., to enable leakage testing), verify which seal is defined as the containment boundary. Ensure that seal grooves are appropriately sized. Verify that the temperature of containment boundary seals will remain within their specified allowable limits under both normal conditions of transport and hypothetical accident conditions.

Confirm that the containment system is securely closed by a positive fastening device that cannot be opened unintentionally or by a pressure that may arise within the package.

4.5.1.2 Codes and Standards

Verify that the containment system is in full compliance with the B&PV Division 3 Code including material, design, fabrication, examination, testing, inspection, and certification. This includes an agreement with an Authorized Inspection Agency to provide inspection and audit services for the Design Owners, Packaging Owners, and Class TP Certificate Holders. The SAR should justify the use of other codes, if appropriate.

4.5.1.3 Special Requirements for Damaged Spent Nuclear Fuel

Review the condition and isotopic composition of the spent nuclear fuel proposed for the packaging. If the contents include damaged fuel, verify that it is canned to facilitate handling and to confine gross fuel particles to a known subcritical volume under normal conditions of transport and hypothetical accident conditions. Ensure that appropriate material specifications and the design/fabrication criteria for the can are justified in the SAR. These specifications and criteria should generally be the same as those for containment or criticality support structures as discussed in Section 2 of this SRP. If a screen-type container is used, an appropriate mesh size should be justified. Containment analysis for aluminum-based spent fuel should be in accordance with WSRC-TR-98-00317.

Spent fuel, with more than 0.74 TBq (20 Ci) of plutonium per package, in the form of debris, particles, loose pellets, or fragmented rods or assemblies, must be packaged in a separate inner container (second containment system) in accordance with 10 CFR 71.63(b). Each containment system must separately meet the requirements of 10 CFR 71.51(a)(1) under normal conditions of transport and 10 CFR 71.51(a)(2) under hypothetical accident conditions. Material specifications and design/fabrication criteria for the inner container should be identical to those of the outer containment. In general, the inner container should also meet all requirements of ANSI N14.5 unless otherwise justified in the SAR (e.g., for periodic and pre-shipment leak testing). Review both containment systems as appropriate.

The determination of the fuel condition should be based, as a minimum, on review of fuel records. Fuel which is known or suspected to be damaged should be visually inspected prior to loading. If the visual inspection indicates no damage greater that a hairline crack or a pinhole leak, the fuel may be considered as undamaged.

4.5.2 Containment under Normal Conditions of Transport

4.5.2.1 Pressurization of Containment Vessel

Verify that the maximum normal operating pressure is consistent with that determined in the Thermal Evaluation section of the SAR. The pressure in the containment vessel should be based on the conditions of the package under normal transport conditions, including temperature, release of gases and volatiles from fuel rod cladding breaches, vaporization of contents, etc. (NRC IN 84-72).

4.5.2.2 Containment Criteria

Detailed guidance on procedures for determining the containment criteria is provided in NUREG/CR-6487.

Confirm that the SNF contents are fully described, including fuel type, fuel amount, percent enrichment, burnup, cool time, decay heat, etc. Confirm that the contents evaluated in the Containment Evaluation section of the SAR are consistent with those presented in the General Information section of the SAR.

Verify that the SAR identifies the constituents which comprise the releasable source term, including radioactive gases, volatiles, and powders. For SNF packages, the releasable source term is composed of crud on the outside of the fuel rod cladding that can become aerosolized, and fuel fines, volatiles, and gases that are released from a fuel rod in the event of a cladding breach. Although the residual contamination on the inside surfaces of the packaging (from previous shipments) typically can be ignored in the determination of the releasable source term, this issue should be addressed. Reasonable bounding values for the effective surface activity density (Ci/cm^2) of the crud on fuel rod clads are based on experimental determinations. A computer code, such as ORIGEN2 (ORNL-CCC-371), is used to identify the radionuclides present for a given percent fuel enrichment, burnup, and cool time. Using the individual A_2 values for the crud, fines, gases, and volatiles individually, the effective A_2 of the releasable source-term mixture can be determined by using the relative release fraction for each contributor and the methods from ANSI N14.5. The release fractions and effective specific activities for the various releasable source term contributors for SNF with an initial enrichment of 3.2%, a burnup of 33,000 MWd/MTIHM, and a cool time of 5 years are given in Table 4-1. The release fractions presented in Table 4-1 are considered bounding and have been developed from reasoned argument and experimental data (NUREG/CR-6487). The SAR should justify release fractions and specific activities, as appropriate.

Based on the mass density, effective specific activity, and effective A_2 of the releasable source term, ensure that the maximum permissible release rate and the maximum permissible leakage rate are calculated in accordance with the containment criteria specified in ANSI N14.5. Verify that the maximum permissible leakage rate under normal transport conditions is converted into a reference air leakage rate under standard leakage test conditions according to ANSI N14.5 and NUREG/CR-6487.

Table 4-1 Release Fractions and Specific Activities for the Contributors to the Releasable Source Term for Packages Designed to Transport Irradiated Fuel Rods.[1,2]

Variable	PWR		BWR	
	Normal conditions of transport	Hypothetical accident conditions	Normal conditions of transport	Hypothetical accident conditions
Fraction of crud that spalls-off of rods, f_C	0.15	1.0	0.15	1.0
Crud surface activity, S_C [Ci/cm^2]	140×10^{-6}	140×10^{-6}	1254×10^{-6}	1254×10^{-6}
Mass fraction of fuel that is released as fines due to a cladding breach, f_F	3×10^{-5}	3×10^{-5}	3×10^{-5}	3×10^{-5}
Specific activity of fuel rods, A_R [Ci/g]	0.60	0.60	0.51	0.51
Fraction of rods that develop cladding breaches, f_B	0.03	1.0	0.03	1.0
Fraction of gases that are released due to a cladding breach, f_G	0.3	0.3	0.3	0.3
Specific activity of gases in a fuel rod, A_G [Ci/g]	7.32×10^{-3}	7.32×10^{-3}	6.28×10^{-3}	6.28×10^{-3}
Specific activity of volatiles in a fuel rod, A_V [Ci/g]	0.1375	0.1375	0.1794	0.1794
Fraction of volatiles that are released due to a cladding breach, f_V	2×10^{-4}	2×10^{-4}	2×10^{-4}	2×10^{-4}

1. 3.2% initial enrichment, 33,000 MWd/MTIHM burnup, 5-year cooling.
2. Applicable only to undamaged fuel. Release fractions for damaged fuel should be justified in the SAR.

Verify that the following maximum permissible leakage rates are determined in accordance with ANSI N14.5:

•   Fabrication verification

•   Periodic verification

•   Maintenance verification

•   Pre-shipment verification.

4.5.2.3 Compliance with Containment Criteria

Confirm that the SAR demonstrates that the package satisfies the containment requirements of 10 CFR 71.51(a)(1) for normal conditions of transport.

- If compliance is demonstrated by test, verify that the leakage rate of a package subjected to the tests of 10 CFR 71.71 does not exceed the temperature- and pressure-corrected air leakage rate.

- If compliance is demonstrated by analysis, verify that the structural evaluation shows that the containment boundary or closure bolts do not undergo any plastic deformation and that the materials of the containment system (e.g., seals) do not exceed their maximum allowable temperature limits when subjected to the conditions in 10 CFR 71.71.

Compliance with the leakage rates for fabrication and periodic verification is discussed in the Acceptance Tests and Maintenance Program Review section of this SRP; compliance with the leakage rates for assembly verification is discussed in the Operating Procedures Review section of this SRP.

4.5.3 Containment under Hypothetical Accident Conditions

The review procedures for containment under hypothetical accident conditions are analogous to those listed in Section 4.5.2 above for normal conditions of transport. Differences relevant to hypothetical accident conditions are noted below.

4.5.3.1 Pressurization of Containment Vessel

The pressure in the containment vessel should be based on the conditions of the package under hypothetical accident conditions, including temperature, release of gases and volatiles from fuel rod cladding breaches, vaporization of contents, etc. Verify that this pressure is consistent with that determined in the Thermal Evaluation section of the SAR.

4.5.3.2 Containment Criteria

The releasable source term, maximum permissible release rate, maximum permissible leakage rate, and conversion to the reference air leakage rate should be based on package conditions and the 10 CFR Part 71 containment requirements under hypothetical accident conditions. Verify that the temperatures, pressure, and physical conditions of the package (including the contents) are consistent with those determined in the Structural Evaluation and Thermal Evaluation sections of the SAR.

Ensure that the reference air leakage rate calculated for hypothetical accident conditions is greater than that determined in Section 4.5.2.2 for normal conditions of transport. In the rare event that this is not the case, the containment criteria for the fabrication, periodic, and assembly verification tests should be based on the hypothetical accident leakage, rather than normal conditions of transport.

The containment requirements of 10 CFR 71.51(a)(2) for hypothetical accident conditions shall be applied individually for krypton-85 and the other radioactive materials. Krypton-85 shall not exceed 10 A_2 in a week. The remaining radioactive materials shall not exceed A_2 in a week.

4.5.3.3 Compliance with Containment Criteria

Confirm that the SAR demonstrates that the package satisfies the containment requirements of
10 CFR 71.51(a)(2) for hypothetical accident conditions. Demonstration is similar to that discussed in
Section 4.5.2.3 above except that the package should be subjected to the tests of 10 CFR 71.73.

4.5.4 Appendix

The appendix may include a list of references, copies of any applicable references not generally available
to the reviewer, computer code descriptions, input and output files, test results, and other appropriate
supplemental information.

4.6 EVALUATION FINDINGS

The containment review should result in the following findings, as appropriate:

4.6.1 Description of Containment System

The staff has reviewed the description and evaluation of the containment system and concludes that: (1)
the SAR identifies established codes and standards for the containment system; (2) the package includes
a containment system securely closed by a positive fastening device that cannot be opened unintentionally
or by a pressure that may arise within the package; (3) the package is made of materials and construction
that assure that there will be no significant chemical, galvanic, or other reaction; (4) a package valve or
similar device, if present, is protected against unauthorized operation and, except for a pressure relief
valve, is provided with an enclosure to retain any leakage; (5) a package designed for the transport of
damaged SNF includes packaging of the damaged SNF in a separate inner container that meets the
requirements of 10 CFR 71.63(c).

4.6.2 Containment under Normal Conditions of Transport

The staff has reviewed the evaluation of the containment system under normal conditions of transport and
concludes that the package is designed, constructed, and prepared for shipment so that under the tests
specified in 10 CFR 71.71 (normal conditions of transport) the package satisfies the containment
requirements of 10 CFR 71.43(f) and 10 CFR 71.51(a)(1) for normal conditions of transport with no
dependence on filters or a mechanical cooling system.

4.6.3 Containment under Hypothetical Accident Conditions

The staff has reviewed the evaluation of the containment system under hypothetical accident conditions
and concludes that the package satisfies the containment requirements of 10 CFR 71.51(a)(2) for
hypothetical accident conditions, with no dependence on filters or a mechanical cooling system.

In summary, the staff has reviewed the Containment Evaluation section of the SAR and concludes that
the package has been described and evaluated to demonstrate that it satisfies the containment
requirements of 10 CFR Part 71, and that the package meets the containment criteria of ANSI N14.5.

4.7 REFERENCES

ANSI N14.5
Institute for Nuclear Materials Management, ANSI N14.5, "American National Standard for Leakage Tests on Packages for Shipment of Radioactive Materials," New York, NY, 1997.

B&PV Division 3 Code
American Society of Mechanical Engineers, "ASME Boiler and Pressure Vessel Code, Section III, Division 3, Containment Systems and Transport Packagings For Spent Nuclear Fuel and High Level Radioactive Waste," New York, NY, 1998.

NRC Bulletin 96-04
U.S. Nuclear Regulatory Commission, "Chemical, Galvanic, or Other Reactions in Spent Fuel Storage and Transportation Casks," OMB No. 3150-0011, Bulletin 96-04, U.S. Government Printing Office, Washington, D.C., July 5, 1996.

NUREG/CR-6487
U.S. Nuclear Regulatory Commission, "Containment Analysis for Type B Packages Used to Transport Various Contents," NUREG/CR-6487, U.S. Government Printing Office, Washington, D.C., 1996.

ORNL-CCC-371
Oak Ridge National Laboratory, "ORIGEN2.1: Isotope Generation and Depletion Code-Matrix Exponential Method," CCC-371, Oak Ridge, TN, 1991.

WSRC-TR-98-00317
Westinghouse Savannah River Company, "Bases for Containment Analysis for Transportation of Aluminum-Based Spent Nuclear Fuel," WSRC-TR-98-00317, Westinghouse Savannah River Company, Aiken, SC, October 1998.

5 SHIELDING REVIEW

5.1 REVIEW OBJECTIVE

The objective of this review is to verify that the package design satisfies the external radiation requirements of 10 CFR Part 71 under normal conditions of transport and hypothetical accident conditions.

5.2 AREAS OF REVIEW

The SAR must be reviewed for adequacy of the description and evaluation of the shielding design. Areas of review include the following:

5.2.1 Description of the Shielding Design
5.2.1.1 Packaging Design Features
5.2.1.2 Codes and Standards
5.2.1.3 Summary Table of Maximum Radiation Levels

5.2.2 Source Specification
5.2.2.1 Gamma Source
5.2.2.2 Neutron Source

5.2.3 Model Specification
5.2.3.1 Configuration of Source and Shielding
5.2.3.2 Material Properties

5.2.4 Evaluation
5.2.4.1 Methods
5.2.4.2 Key Input and Output Data
5.2.4.3 Flux-to-Dose-Rate Conversion
5.2.4.4 Radiation Levels

5.2.5 Appendix

5.3 REGULATORY REQUIREMENTS

Regulatory requirements of 10 CFR Part 71 applicable to the shielding review are as follows:

5.3.1 Description of the Shielding Design

The packaging must be described in sufficient detail to provide an adequate basis for its evaluation. The description of the shielding components must include dimensions, materials of construction, and materials specifically used for neutron shielding. [10 CFR 71.31(a)(1) and 10 CFR 71.33(a)(5)]

The SAR must identify established codes and standards applicable to the shielding design. [10 CFR 71.31(c)]

5.3.2 Source Specification

The contents must be described in sufficient detail to provide an adequate basis for their evaluation. This description must include those radionuclides that result in the largest external radiation levels, and their chemical and physical form. [10 CFR 71.31(a)(1), 10 CFR 71.33(b)(1), 10 CFR 71.33(b)(2), and 10 CFR 71.33(b)(3)]

5.3.3 Model Specification

The package must be described and evaluated to demonstrate that it satisfies the shielding requirements of 10 CFR Part 71. [10 CFR 71.31(a) and 10 CFR 71.31(b)]

5.3.4 Evaluation

The package must be evaluated to demonstrate that it satisfies the shielding requirements specified in 10 CFR Part 71, Subpart E. [10 CFR 71.31(a)(2), 10 CFR 71.35(a), and 10 CFR 71.41(a)]

The package must be designed, constructed, and prepared for shipment so that the external surface radiation levels will not significantly increase under the tests specified in 10 CFR 71.71 (normal conditions of transport). [10 CFR 71.43(f) and 10 CFR 71.51(a)(1)]

Under the tests specified in 10 CFR 71.71 (normal conditions of transport), the external radiation levels must satisfy the requirements of 10 CFR 71.47(b) for exclusive-use shipments. (10 CFR 71.47(a) requirements for non-exclusive-use shipments are assumed not to apply to spent nuclear fuel (SNF) packages.) The package and vehicle radiation limits for exclusive-use shipments are summarized in Table 5-1.

Under the tests specified in 10 CFR 71.73 (hypothetical accident conditions), the external radiation levels at 1 m (40 in) from the package surface must not exceed 10 mSv/hr (1 rem/hr). [10 CFR 71.51(a)(2)]

5.4 ACCEPTANCE CRITERIA

5.4.1 Description of the Shielding Design

The regulatory requirements in Section 5.3.1 identify the acceptance criteria.

5.4.2 Source Specification

The regulatory requirements in Section 5.3.2 identify the acceptance criteria.

5.4.3 Model Specification

In addition to the regulatory requirements identified in Section 5.3.3, the model used in the shielding evaluation should be described in sufficient detail to permit an independent review, with confirmatory calculations, of the package shielding design.

Table 5-1 External Radiation Level Limits for Exclusive-Use Shipments

Package and Vehicle Radiation Level Limits (49 CFR 173.441, 10 CFR 71.47(b)) This table must not be used as a substitute for DOT or NRC regulations on transportation of radioactive materials.			
Transport Vehicle Use:	Exclusive		
Transport Vehicle Type:	Open (flat-bed)	Open w/Enclosure[a]	Closed
Package (or freight container) **Limits:**			
External Surface	2 mSv/hr **(200 mrem/hr)**	10 mSv/hr **(1000 mrem/hr)**	10 mSv/hr **(1000 mrem/hr)**
Roadway or Railway Vehicle (or freight container) **Limits:**			
Any point on the outer surface	N/A	N/A	2 mSv/hr **(200 mrem/hr)**
Vertical planes projected from outer edges	2 mSv/hr **(200 mrem/hr)**	2 mSv/hr **(200 mrem/hr)**	N/A
Top of . . .	load: 2 mSv/hr **(200 mrem/hr)**	enclosure: 2 mSv/hr **(200 mrem/hr)**	vehicle: 2 mSv/hr **(200 mrem/hr)**
2 m (80 in) from. . .	vertical planes: 0.1 mSv/hr **(10 mrem/hr)**	vertical planes: 0.1 mSv/hr **(10 mrem/hr)**	outer lateral surfaces: 0.1 mSv/hr **(10 mrem/hr)**
Underside	2 mSv/hr **(200 mrem/hr)**		
Occupied position	0.02 mSv/hr **(2 mrem/hr)**[b]		

a. Securely attached (to vehicle), access-limiting enclosure; package personnel barriers are considered as enclosures.
b. Does not apply to private carrier wearing dosimetry if under radiation protection program satisfying 10 CFR Part 20.

5.4.4 Evaluation

In addition to the regulatory requirements identified in Section 5.3.4 and Table 5-1, shielding should not exceed its allowable temperature limits under normal conditions of transport or hypothetical accident conditions.

5.5 REVIEW PROCEDURES

The following procedures are generally applicable to the shielding review of SNF transportation packages. Since packages for shipment of SNF are generally intended to be shipped by exclusive-use, only exclusive-use shipments are assumed in the following SRP review procedures.

The shielding review is based in part on the descriptions and evaluations presented in the General Information, Structural Evaluation, and Thermal Evaluation sections of the SAR. Similarly, results of the shielding review are considered in the review of the SAR sections on Operating Procedures and Acceptance Tests and Maintenance Program. Examples of SAR information flow into, within, and from the shielding review are shown in Figure 5-1.

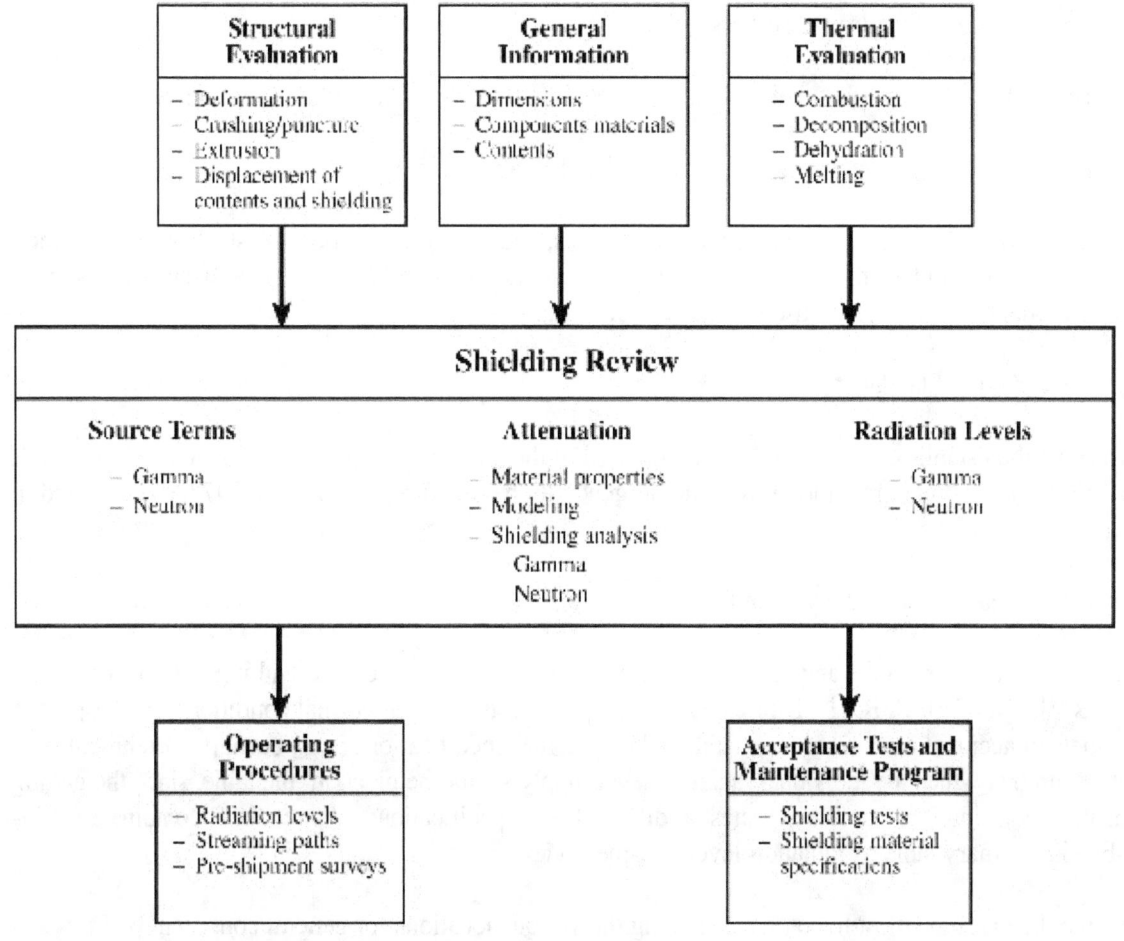

Figure 5-1 SAR Information Flow for the Shielding Review.

5.5.1 Description of the Shielding Design

5.5.1.1 Packaging Design Features

Review the General Information chapter of the SAR and any additional description of the shielding design presented in the Shielding Evaluation section. Design features important to safety include, but are not limited to:

- Dimensions, tolerances, and densities of material for neutron and gamma shielding, including those of structural or thermal components considered in the shielding evaluation

- Concentrations of neutron absorbers

- Structural components that maintain the contents in a fixed position within the package

- Dimensions of the conveyance that are considered in the shielding evaluation.

All information presented in the text, drawings, figures, and tables should be consistent with each other and with that used in the shielding evaluation. Pay close attention to consistency between the drawings, and the models and parameters used in the shielding analysis.

5.5.1.2 Codes and Standards

Verify that the established codes and standards used in the shielding design are identified. For example, conversion of the flux to radiation levels should generally be based on ANSI 6.1.1-1977, as discussed in Section 5.5.4.3.

5.5.1.3 Summary Table of Maximum Radiation Levels

Examine the summary table, and verify that the maximum radiation levels are within the limits of 10 CFR 71.47 and 10 CFR 71.51 for exclusive-use shipments for both normal conditions of transport and hypothetical accident conditions (see Table 5-1). The fuel specifications (e.g., burnup, enrichment, cooling time) at which the individual radiation levels apply should be given in the table, since the gamma or neutron contributions could be greatest at different fuel specifications. Appendix B contains a fill-in-the-blank summary table of radiation levels for the reviewer's use.

Examine the variation of radiation levels among the various locations for general consistency. For example, radiation levels should decrease with increased distance from the source or greater shielding effectiveness.

5.5.2 Source Specification

Compare the specifications for the SNF contents with those listed in the General Information section of the SAR. The ranges of fuel type, burnup, enrichment, and cooling time should be stated. Generally, the General Information section will specify a maximum fuel enrichment, which is important in the criticality analysis. For shielding evaluations, however, the neutron source term increases considerably with decreasing initial enrichment at a constant burnup. Consequently, the SAR should also specify a minimum initial enrichment.

Generally, the applicant will use a computer code such as ORIGEN-S (NUREG/CR-0200A) [or a SAS2 sequence of SCALE], ORIGEN2 (Oak Ridge 1991), or the Department of Energy Characteristics Data Base (TRW-CSCIID A00020002-AAX01.1) to determine the source terms. The latter two have energy group structure limitations that are discussed below. If the applicant has chosen ORIGEN2, verify that the cross-section library is appropriate for the fuel being considered. Many libraries are not appropriate for burnup that exceeds 33,000 MWd/MTU.

5.5.2.1 Gamma Source

Verify that the gamma source terms are specified as a function of energy for both the SNF and activated hardware. If the energy group structure of the source term calculation differs from that of the cross section set of the shielding calculation, the applicant may need to regroup the photons. One method of regrouping is to input the nuclide activities from the source term calculation to a simple decay code with a variable group structure (e.g., GAMGEN [Gosnell 1990]). In general, only gammas from approximately 0.8 to 2.5 MeV will contribute significantly to the external radiation levels, so regrouping outside of this range is of little consequence. Pay particular attention to whether the source terms are specified per assembly, per total number of assemblies, or per metric ton, and ensure that the total source is correctly used in the shielding calculation.

Determining the source terms for fuel assembly hardware is generally not as straightforward as that for the SNF, especially if one of the ORIGEN codes is used. The activation of the hardware depends on the impurities (e.g., ^{59}Co) initially present and on the spatial and energy variation of the neutron flux during burnup. The effort devoted to reviewing the calculation of the hardware source term should be appropriate to its contribution to the radiation levels presented in the shielding analysis. Note also whether the package is intended to transport other hardware such as control assemblies or shrouds, and ensure that the source terms from these components are also included if applicable. Two reports that may be helpful in reviewing the calculation of hardware activation are ORNL/TM-11018 and PNL-6906.

Depending on the packaging design, neutron interactions could result in the production of energetic gammas near the packaging surface. If this source is not treated by the shielding analysis code, verify that it is determined by other appropriate means.

The result of the source term calculation should be a listing of gammas per second, or MeV per second, as a function of energy. The activity of each nuclide that contributes significantly to the source term should be provided as supporting information.

Because the gamma radiation levels are directly proportional to the gamma source term, the review should independently confirm the source term calculated by the applicant.

5.5.2.2 Neutron Source

Verify that the neutron source term is expressed as a function of energy. The neutron source will generally result from both spontaneous fission of transuranics and from (a,n) reactions in the fuel. Depending on the methods used to calculate these source terms, the applicant might determine the energy group structure independently. This is often accomplished by selecting the nuclide with the predominant contribution to spontaneous fission (e.g., ^{244}Cm) and using that spectrum for all neutrons, since the contribution from (a,n) reactions is generally small. Assure that neutron multiplication in the fissile material is included in the analysis. The fissile content assumed for the multiplication effect should be justified and conservative.

The result of the source term calculation or experimental data should be a listing of neutrons per second as a function of energy. For the spontaneous fission contribution, a listing of the significant nuclides should also be presented.

Because the neutron radiation levels are directly proportional to the neutron source term, the review should independently confirm the source term calculated by the applicant.

5.5.3 Model Specification

Review the Structural Evaluation and Thermal Evaluation sections of the SAR to determine the effects of the tests for both normal conditions of transport and hypothetical accident conditions on the packaging and its contents. For example, the package might have impact limiters or an external neutron shield that would be damaged or destroyed during the structural and thermal tests of 10 CFR 71.73. Verify that the corresponding models used in the shielding calculation are consistent with these effects.

5.5.3.1 Configuration of Source and Shielding

Examine the sketches or figures that indicate how the shielding is modeled. Verify that the model dimensions and materials are consistent with those specified in the package drawings presented in the General Information section of the SAR and the normal and accident conditions of the package. Dimensions should be at the conservative end of their tolerance range. Ensure that voids, streaming paths, and irregular geometries are taken into account or otherwise modeled conservatively. Differences between normal conditions of transport and hypothetical accident conditions should be clearly indicated.

Verify that the source term locations for both SNF and the structural support regions of the fuel assemblies are modeled properly. Generally, at least three source regions (fuel and top/bottom assembly hardware) are necessary. Within the SNF region, the fuel materials may generally be homogenized to facilitate shielding calculations. In some cases, the basket material may be homogenized also. The reviewer should watch for cases when homogenization is not appropriate, such as when it distorts the neutron multiplication rate or when radiation streaming can occur between the basket components.

Because of the burnup profile, a uniform source distribution is generally conservative for the top and bottom dose points, but not for the axial center unless the source strength is appropriately adjusted. If peaking appears to be significant, verify that it has been treated appropriately. The assembly structural support regions (e.g., top/bottom end-pieces and plenum) should be correctly positioned relative to the SNF. These support regions may be individually homogenized.

If transport is by exclusive use (as it typically is for SNF), dimensions of the transport vehicle should be included, as appropriate (e.g., to determine the radiation level at 2 m from the vehicle). If the vehicle occupants do not wear dosimetry devices under a radiation protection program in conformance with 10 CFR 20.1502, applicable vehicle dimensions will also be necessary to determine the radiation level at normally occupied locations. (These dimensions or vehicle type, as well as positioning of the packages, may become limiting conditions in the certificate of compliance for exclusive-use shipments.)

Verify that the dose point locations for the various calculations include all locations prescribed in 10 CFR 71.47(b) and 10 CFR 71.51(a)(2). Ensure that the dose points are chosen to identify the location of the maximum radiation levels; the maximum might not occur at the midpoint of a package surface or parallel plane. Radiation peaking often occurs near the edges of external neutron shields and impact limiters. Determine that voids, streaming paths, and irregular geometries are included in the model or otherwise treated in a conservative manner.

5.5.3.2 Material Properties

Verify that the mass densities and atom densities are provided for all materials used in the models of the packaging, source, and conveyance (if applicable). Because most computer programs for shielding calculations now allow input in either g/cm^3 or atoms/barn-cm, the review may consider either mass or atom densities alone to be sufficient for certain materials. Atom densities are subject to frequent error and should be confirmed if used as input to shielding calculations. For uncommon materials, especially foams, plastics, and other hydrocarbons, the source of the data should be referenced. Ensure that these materials are properly controlled to achieve such densities. (Specific information on control measures should be included in the Acceptance Tests and Maintenance Program section of the SAR.) Review materials to assess if any shielding properties could degrade during the service life of the packaging. Confirm that controls are in place to ensure the long term effectiveness of the shielding, as appropriate.

Confirm that temperature-sensitive shielding materials will not be subject to temperatures at or above their design limitations during either normal or accident conditions. Determine whether the applicant properly examined the potential for shielding material to experience changes in material densities at temperature extremes. (For example, elevated temperatures may reduce hydrogen content through loss of bound or free water in hydrogenous shielding materials.)

As noted above, a common practice in shielding analyses is to homogenize the source region rather than develop a detailed heterogeneous model of every fuel pin, pellet, or similar contents. Because an accurate effective density of the homogenized source is important for self-shielding, a confirmatory calculation of this density is generally warranted.

5.5.4 Evaluation

5.5.4.1 Methods

Verify that the computer program(s) used for the shielding analysis are appropriate. These codes may use Monte Carlo transport, deterministic transport, or point-kernel techniques for problem solution. (The latter is generally appropriate only for gammas since transportation packagings typically do not contain sufficient hydrogenous material to apply removal cross sections for point-kernel neutron calculations.) Shielding codes that are typically used in SARs include, but are not limited to, TORT/DORT (ORNL-6268), DANTSYS (LA-9184-M and LA-I 0049-M), MCNP (LA-12625-M), COG (Buck 1994), and MORSE (NUREG/CR-0200B) or one of the SAS sequence codes in SCALE. For computer codes not well established in the public domain, the SAR should describe the solution method, benchmark results, validation procedures, and quality assurance practices.

Assess if the number of dimensions of the code is appropriate for the cask configuration. Generally, a 2-D or 3-D calculation is necessary. One-dimensional codes provide little information about off-axis locations and streaming paths. Even for radiation levels at the end of the package, 1-D codes require a buckling correction that must be justified; merely using the packaging cavity diameter may underestimate the radiation level (overestimate the radial leakage).

Verify that the cross section library used by the code is applicable for shielding calculations. Ensure that a coupled cross section set is used and that the code has been executed in a manner that accounts for secondary source terms, unless the evaluation has independently determined a source term for neutron-induced gamma radiation or subcritical multiplication of neutrons.

5.5.4.2 Key Input and Output Data

Verify that key input data for the shielding calculations are identified. These will depend on the type of code (point kernel, deterministic, Monte Carlo, etc.) as well as the code itself. In addition to the source terms, materials, and dimensions identified above, key input data can include convergence criteria, mesh size, neutrons per generation, number of generations, etc. Verify that the information from the shielding model is properly input into the code.

The SAR should also generally include a representative output file (or key sections of the file including input data) for each type of calculation performed in the shielding analysis. The review should ensure that proper convergence is achieved and that the calculated radiation levels from the output files agree with those reported in the text.

5.5.4.3 Flux-to-Dose-Rate Conversion

The shielding analysis code will typically have the capability to perform this conversion directly with its own data library or with one supplied by the user. Generally this conversion will use ANSI/ANS 6.1.1-1977.

Verify the accuracy of the flux-to-dose-rate conversion factors, which should be tabulated as a function of the energy group structure used in the shielding calculation.

5.5.4.4 Radiation Levels

Confirm that the radiation levels under both normal conditions of transport and hypothetical accident conditions are in agreement with the summary tables in Section 5.5.1.3 above and that they satisfy the limits in 10 CFR 71.47(b) and 10 CFR 71.51(a)(2). Verify that the analysis shows that the locations selected are those of maximum radiation levels and include any radiation streaming paths.

For the purposes of 10 CFR 71.47(b), NRC staff considers the external surface to be that part of the package which is shown in the drawings and has been demonstrated to remain in place under the tests in 10 CFR 71.71 (normal conditions of transport). Personnel barriers and similar devices that are attached to the conveyance, rather than the package, can, however, qualify the vehicle as a closed vehicle (NUREG/CR-5569A and NUREG/CR-5569B) as defined in 49 CFR 173.403.

Determine that the radiation levels appear reasonable and that their variation with location are consistent with the geometry and shielding characteristics of the package.

Ensure that the evaluation addresses damage to the shielding under normal conditions of transport and hypothetical accident conditions. Verify that any damage under normal conditions of transport (10 CFR 71.71) does not result in a significant increase in external radiation levels, as required by 10 CFR 71.43(f) and 10 CFR 71.51(a)(1). Note that external neutron shielding may not be designed to remain in place under hypothetical accident conditions.

The review may include a confirmatory analysis of the shielding calculations reported in the SAR. Because measurements of the actual radiation levels from packages must be performed prior to shipment in order to show that the 10 CFR 71.47 limits are satisfied, a number of factors should be considered in determining the level of effort of the confirmatory analysis. These factors include such items as the expected magnitude of the radiation levels, similarity with previously reviewed packages, thoroughness of the review of source terms and other input data, bounding assumptions in the analysis, margin from the regulatory limits, and the contribution from difficult to measure neutrons.

At a minimum, the review should include examination of the applicant's input to the computer program used for the shielding analysis. Verify use of proper dimensions, material properties, and an appropriate cross section set. In addition, independently evaluate the use of gamma and neutron source terms.

If a more detailed review is required, independently evaluate the radiation levels to ensure that the SAR results are reasonable and conservative. As previously noted, the use of a simple code for neutron calculations is often not appropriate. An extensive evaluation is necessary if major errors are suspected. To the degree possible, the use of a different shielding code with a different analytical technique and cross section set from that of the SAR analysis will provide a more independent evaluation.

5.5.5 Appendix

The appendix may include a list of references, copies of applicable references if not generally available to the reviewer, computer code descriptions, input and output files, test results, and other appropriate supplemental information.

5.6 EVALUATION FINDINGS

The shielding review should result in the following findings, as appropriate:

5.6.1 Description of the Shielding Design

The staff has reviewed the package description and evaluation and concludes that they satisfy the shielding requirements of 10 CFR Part 71.

5.6.2 Source Specification

The staff has reviewed the source specification used in the shielding evaluation and concludes that they are sufficient to provide a basis for evaluation of the package against the 10 CFR Part 71 shielding requirements.

5.6.3 Model Specification

The staff has reviewed the models used in the shielding evaluation and concludes that they are described in sufficient detail to permit an independent review, with confirmatory calculations, of the package shielding design.

5.6.4 Evaluation

The staff has reviewed the external radiation levels of the package and vehicle as it will be prepared for shipment and concludes that they satisfy 10 CFR 71.47(b) for packages transported by exclusive-use vehicle.

The staff has reviewed the package design, construction, and preparations for shipment and concludes that the external radiation levels will not significantly increase during normal conditions of transport consistent with the tests specified in 10 CFR 71.71.

The staff has reviewed the package design, construction, and preparations for shipment and concludes that the maximum external radiation level at one meter from the external surface of the package will not exceed 10 mSv/hr (1 rem/hr) during hypothetical accident conditions consistent with the tests specified in 10 CFR 71.73.

5.7 REFERENCES

ANSI/ANS 6.1.1-1977   American Nuclear Society, ANSI/ANS 6.1.1, "American National Standard for Neutron and Gamma-Ray Flux to Dose Factors," La Grange Park, IL, 1977.

ANSI/ANS 6.1.1-1991   American Nuclear Society, ANSI/ANS 6.1.1, "American National Standard for Neutron and Gamma-Ray Fluence to Dose Factors," La Grange Park, IL, 1991.

Buck 1994

Buck, R.M. et al., "COG-A Monte Carlo Neutron, Photon, Electron Transport Code," Lawrence Livermore National Laboratory, M-221-1, Livermore, CA July 4, 1994.

60 FR 50247

U.S. Nuclear Regulatory Commission, "Compatibility With the International Atomic Energy Agency (IAEA)," *Federal Register*, FR 50247, U.S. Government Printing Office, Washington, D.C., September 28, 1995.

Gosnell 1990

Gosnell, T.B., "Automated Calculation of Photon Source Emission From Arbitrary Mixtures of Naturally Radioactive Heavy Nuclides," in Editor, *Nuclear Instruments and Methods in Physics Research,* **A299** (1990), Elsevier Science, Elmont, NY, pp. 682-686.

LA-I 0049-M

Los Alamos National Laboratory, "User's Guide for TWODANT: A Code Package for Two Dimensional, Diffusion Accelerated Neutral Particle Transport," LA-I 0049-M Rev., Los Alamos, NM, April 1992.

LA-9184-M

Los Alamos National Laboratory, "Revised User's Manual for ONEDANT: A Code for One Dimensional, Diffusion Accelerated Neutral Particle Transport," LA-9184-M, Rev., Los Alamos, NM, December 1989.

LA-12625-M

Los Alamos National Laboratory, "MCNP 4A, Monte Carlo N-Particle Transport Code System," LA-12625-M, Los Alamos, NM, December 1993.

NRC IN 80-32

U.S. Nuclear Regulatory Commission, "Clarification of Certain Requirements for Exclusive-Use Shipments of Radioactive Materials," IE Information Notice 80-32, U.S. Government Printing Office, Washington, D.C., August 29, 1980.

NUREG/CR-0200A

U.S. Nuclear Regulatory Commission, "SCALE: A Modular Code System for Performing Standardized Computer Analyses for Licensing Evaluation," NUREG/CR-0200, Vol. 2, Part 1, Rev. 4, U.S. Government Printing Office, Washington, D.C., April 1995.

NUREG/CR-0200B

U.S. Nuclear Regulatory Commission, "SCALE: A Modular Code System for Performing Standardized Computer Analyses for Licensing Evaluation," NUREG/CR-0200, Vol. 2, Part 2, Rev. 4, U.S. Government Printing Office, Washington, D.C., April 1995.

NUREG/CR-5569A

U.S. Nuclear Regulatory Commission, "Clarification of Certain Requirements for Exclusive-Use Shipments of Radioactive Materials," HPPOS-084, in *Health Physics Positions Data Base,* NUREG/CR-

5569, Rev. 1, U.S. Government Printing Office, Washington, D.C., February 1991.

NUREG/CR-5569B   U.S. Nuclear Regulatory Commission, "Clarification of Certain Requirements for Exclusive-Use Shipments," HPPOS-085, in *Health Physics Positions Data Base*, NUREG/CR-5569, Rev. 1, U.S. Government Printing Office, Washington, D.C., February 1991.

ORNL-CCC-371   Oak Ridge National Laboratory, "ORIGEN2.1: Isotope Generation and Depletion Code-Matrix Exponential Method," CCC-371, Oak Ridge, TN, 1991.

ORNL-6268   Oak Ridge National Laboratory, "The TORT Three-Dimensional Discrete Ordinates Neutron/Photon Transport Code," ORNL-6268, Oak Ridge, TN, November 1987.

ORNL/TM-11018   Ludwig, S.B., and Renier, J.P., "Standard- and Extended-Burnup PWR and BWR Reactor Models for the ORIGEN2 Computer Code," ORNL/TM-11018, Oak Ridge National Laboratory, Oak Ridge, TN, December 1989.

PNL-6906   Luksic, A., "Spent Fuel Assembly Hardware Characterization and 10 CFR 61 Classification for Waste Disposal," PNL-6906, Volume 1, Pacific Northwest Laboratory, Richland, WA, June 1989.

TRW- CSCIID   TRW Environmental Safety Systems, Inc., "DOE Characteristics Data
A00020002-AAX01.0   Base, User Manual for the CDB_R," CSCIID A00020002-AAX01.0, Vienna, VA, November 16, 1992.

6 CRITICALITY REVIEW

6.1 REVIEW OBJECTIVE

The objective of this review is to verify that the package design satisfies the criticality safety requirements of 10 CFR Part 71 under normal conditions of transport and hypothetical accident conditions.

6.2 AREAS OF REVIEW

The SAR should be reviewed for adequacy of the description and evaluation of the criticality design. Areas of review include the following:

6.2.1 Description of Criticality Design
6.2.1.1 Packaging Design Features
6.2.1.2 Codes and Standards
6.2.1.3 Summary Table of Criticality Evaluations
6.2.1.4 Transport Index

6.2.2 Spent Nuclear Fuel Contents

6.2.3 General Considerations for Criticality Evaluations
6.2.3.1 Model Configuration
6.2.3.2 Material Properties
6.2.3.3 Computer Codes and Cross Section Libraries
6.2.3.4 Demonstration of Maximum Reactivity
6.2.3.5 Confirmatory Analyses

6.2.4 Single Package Evaluation
6.2.4.1 Configuration
6.2.4.2 Results

6.2.5 Evaluation of Package Arrays under Normal Conditions of Transport
6.2.5.1 Configuration

6.3 REGULATORY REQUIREMENTS

Regulatory requirements of 10 CFR Part 71 applicable to the criticality review are as follows:

6.3.1 Description of Criticality Design

The packaging must be described in sufficient detail to provide an adequate basis for its evaluation. This description must include types and dimensions of materials of construction and materials specifically used as nonfissile neutron absorbers or moderators. [10 CFR 71.31(a)(1) and 10 CFR 71.33(a)(5)]

The SAR must identify established codes and standards applicable to the criticality design. [10 CFR 71.31(c)]

The SAR must specify the allowable number of packages that may be transported in a single shipment. [10 CFR 71.35(b)]

A fissile material package must be assigned a transport index for nuclear criticality control. [10 CFR 71.59(b)]

6.3.2 Spent Nuclear Fuel Contents

The contents must be described in sufficient detail to provide an adequate basis for their evaluation. This description must include the type, maximum quantity, and chemical and physical form of the spent nuclear fuel (SNF). [10 CFR 71.31(a)(1), 10 CFR 71.33(b)(1), 10 CFR 71.33(b)(2), and 10 CFR 71.33(b)(3)]

Unknown properties of fissile material must be assumed to be those which will result in the highest neutron multiplication. [10 CFR 71.83]

6.3.3 General Considerations for Criticality Evaluations

The package must be evaluated to demonstrate that it satisfies the criticality safety requirements of 10 CFR Part 71, Subpart E. [10 CFR 71.31(a)(2), 10 CFR 71.35(a), and 10 CFR 71.41(a)]

6.3.4 Single Package Evaluation

A single package must satisfy the specifications of 10 CFR 71.43(f), 10 CFR 71.51(a)(1), and 10 CFR 71.55(d) under normal conditions of transport. These requirements address subcriticality, alteration of the geometric form of the contents, inleakage of water, and effectiveness of the packaging. [10 CFR 71.35, 10 CFR 71.43(f), 10 CFR 71.51(a)(1), and 10 CFR 71.55(d)]

A single package must be designed and constructed and its contents limited so that it would be subcritical if water were to leak into the containment system. [10 CFR 71.55(b)]

A single package must be subcritical under the tests for hypothetical accident conditions. [10 CFR 71.55(e)]

6.3.5 Evaluation of Package Arrays under Normal Conditions of Transport

The SAR must evaluate arrays of packages under normal conditions of transport to determine the maximum number of packages that may be transported in a single shipment. [10 CFR 71.35 and 10 CFR 71.59]

6.3.6 Evaluation of Package Arrays under Hypothetical Accident Conditions

The SAR must evaluate arrays of packages under hypothetical accident conditions to determine the maximum number of packages that may be transported in a single shipment. [10 CFR 71.35 and 10 CFR 71.59]

6.3.7 Benchmark Evaluations

The package must be evaluated to demonstrate that it satisfies the criticality safety requirements of 10 CFR Part 71. [10 CFR 71.31(a)(2) and 10 CFR 71.35]

6.3.8 Burnup Credit

There are no regulatory requirements that are specific to burnup credit. The general criticality requirements apply. However, based on experience, the staff has developed guidelines to facilitate the review of burnup credit, when it is included in the analysis. Burnup credit evaluations are performed in accordance with Sections 6.4.8.1 through 6.4.8.6.

6.4 ACCEPTANCE CRITERIA

6.4.1 Description of Criticality Design

The regulatory requirements in Section 6.3.1 identify the acceptance criteria.

6.4.2 Spent Nuclear Fuel Contents

The regulatory requirements in Section 6.3.2 identify the acceptance criteria.

6.4.3 General Considerations for Criticality Evaluations

In addition to the regulatory requirements identified in Section 6.3.3, the packaging model for the criticality evaluation should generally consider no more than 75% of the specified minimum neutron poison concentrations. The model for the SNF should include no burnable poisons. Methods for including fuel burnup in the criticality calculations need to have prior approval by NRC.

The sum of the effective multiplication factor (k_{eff}), two standard deviations (95% confidence), and the bias adjustment should not exceed 0.95 to demonstrate subcriticality by calculation. A bias that reduces the calculated value of k_{eff} should not be applied.

6.4.4 Single Package Evaluation

In addition to the regulatory requirements identified in Section 6.3.4, the assumption of water inleakage for the analysis pursuant to 10 CFR 71.55 (b) should consider the packaging and contents to be in their most reactive condition, as determined by the tests in 10 CFR 71.71 and 10 CFR 71.73.

6.4.5 Evaluation of Package Arrays under Normal Conditions of Transport

The regulatory requirements in Section 6.3.5 identify the acceptance criteria.

6.4.6 Evaluation of Package Arrays under Hypothetical Accident Conditions

The regulatory requirements in Section 6.3.6 identify the acceptance criteria.

6.4.7 Benchmark Evaluations

The criticality evaluation of the package should include a comparison of the calculational methods with applicable benchmark experiments to determine the appropriate bias and uncertainties.

6.4.8 Burnup Credit Evaluation

The staff guidance in Sections 6.4.8.1 through 6.4.8.6 identify the acceptance criteria.

6.4.8.1 Limits for the Licensing Basis

Verify that the licensing-basis analysis performed to demonstrate criticality safety limits the amount of burnup credit to that available from actinide compositions associated with PWR irradiation of UO_2 fuel to an assembly-average burnup value of 40 GWd/MTU or less. This licensing-basis analysis should assume an out-of-reactor cooling time of five years and should be restricted to intact assemblies that have not used burnable absorbers. The initial enrichment of the fuel assumed for the licensing-basis analysis should be no more than 4.0 wt% ^{235}U unless a loading offset is applied. The loading offset is defined as the minimum amount by which the assigned burnup loading value (see Section 6.4.8.5) must exceed the burnup value used in the licensing safety basis analysis. The loading offset should be at least 1 GWd/MTU for every 0.1 wt% increase in initial enrichment above 4.0 wt%. In any case, the initial enrichment shall not exceed 5.0 wt%. For example, if the applicant performs a safety analysis that demonstrates an appropriate subcritical margin for 4.5 wt% fuel burned to the limit of 40 GWd/MTU, then the loading curve (see Section 6.4.8.4) should be developed to ensure that the assigned burnup loading value is at least 45 GWd/MTU (i.e., a 5 GWd/MTU loading offset resulting from the 0.5 wt% excess enrichment over 4.0 wt%). Applicants requesting use of actinide compositions associated with fuel assemblies, burnup values, or cooling times outside these specifications, or applicants requesting a relaxation of the loading offset for initial enrichments between 4.0 and 5.0 wt%, should provide the measurement data and/or justify extrapolation techniques necessary to adequately extend the isotopic validation and quantify or bound the bias and uncertainty.

6.4.8.2.   Code Validation

Ensure that the analysis methodologies used for predicting the actinide compositions and determining the neutron multiplication factor (k-effective) are properly validated. Bias and uncertainties associated with predicting the actinide compositions should be determined from benchmarks of applicable fuel assay measurements. Bias and uncertainties associated with the calculation of k-effective should be derived from benchmark experiments that represent important features of the cask design and spent fuel contents. The particular set of nuclides used to determine the k-effective value should be limited to that established in the validation process. The bias and uncertainties should be applied in a way that ensures conservatism in the licensing safety analysis. Particular consideration should be given to bias uncertainties arising from the lack of critical experiments that are highly prototypical of spent fuel in a cask.

6.4.8.3 Licensing-Basis Model Assumptions.

Ensure that the actinide compositions used in analyzing the licensing safety basis (as described in 6.4.8.1) are calculated using fuel design and in-reactor operating parameters selected to provide conservative estimates of the k-effective value under cask conditions. The calculation of the k-effective value should be performed using cask models, appropriate analysis assumptions, and code inputs that allow adequate representation of the physics. Of particular concern should be the need to account for the axial and horizontal variation of the burnup within a spent fuel assembly (e.g., the assumed axial burnup profiles), the need to consider the more reactive actinide compositions of fuels burned with fixed absorbers or with control rods fully or partly inserted, and the need for a k-effective model that accurately accounts for local reactivity effects at the less-burned axial ends of the fuel region.

6.4.8.4 Loading Curve

Verify that the application includes one or more loading curves that plot, as a function of initial enrichment, the assigned burnup loading value above which fuel assemblies may be loaded in the cask. Loading curves should be established based on a 5-year cooling time and only fuel cooled at least five years should be loaded in a cask approved for burnup credit.

6.4.8.5 Assigned Burnup Loading Value

Verify that administrative procedures are adequately described to ensure that licensees load the cask with fuel that is within the specifications of the approved contents. The administrative procedures should include an assembly measurement that confirms the reactor record assembly burnup. The measurement technique may be calibrated to the reactor records for a representative set of assemblies. For an assembly reactor burnup record to be confirmed, the measurement should provide agreement within a 95 percent confidence interval based on the measurement uncertainty. The assembly burnup value to be used for loading acceptance (termed the assigned burnup loading value) should be the confirmed reactor record value as adjusted by reducing the record value by the combined uncertainties in the records and the measurement.

6.4.8.6 Estimate of Additional Reactivity Margin

Ensure that design-specific analyses are provided that estimate the additional reactivity margins available from fission product and actinide nuclides not included in the licensing safety basis (as described in Section 6.4.8.1). The analysis methods used for determining these estimated reactivity margins should be verified using available experimental data (e.g., isotopic assay data) and computational benchmarks that demonstrate the performance of the applicant's methods in comparison with independent methods and analyses. The Organization for Economic Cooperation and Development Nuclear Energy Agency's Working Group on Burnup Credit provides a source of computational benchmarks that may be considered. The design-specific margins should be evaluated over the full range of initial enrichments and burnups on the burnup credit loading curve(s). The resulting estimated margins should then be assessed against estimates of: (a) any uncertainties not directly evaluated in the modeling or validation processes for actinide-only burnup credit (e.g., k-effective validation uncertainties caused by a lack of critical experiment benchmarks with either actinide compositions that match those in spent fuel or material geometries that represent the most reactive ends of spent fuel in casks); and (b) any potential nonconservatisms in the models for calculating the licensing-basis actinide inventories (e.g., any outlier assemblies with higher-than-modeled reactivity caused by the use of control rod insertion during burnup).

6.5 REVIEW PROCEDURES

The following procedures are generally applicable to the criticality review of SNF transportation packages. Since packages for shipment of SNF are generally intended to be shipped by exclusive-use, only exclusive-use shipments are assumed in the following SRP review procedures.

The criticality review is based in part on the descriptions and evaluations presented in the General Information, Structural Evaluation, and Thermal Evaluation sections of the SAR. Similarly, results of the criticality review are considered in the review of the SAR sections on Operating Procedures and Acceptance Tests and Maintenance Program. Examples of SAR information flow into, within, and from the criticality review are shown in Figure 6-1.

6.5.1 Description of the Criticality Design

6.5.1.1 Packaging Design Features

Review the General Information section of the SAR and any additional description of the criticality design presented in the Criticality Evaluation section. Packaging design features important for criticality include, but are not limited to:

- Dimensions and tolerances of the containment system

- Dimensions, material composition, and tolerances of structural components (e.g., basket) that maintain the SNF in a fixed position within the package or in a fixed position relative to neutron absorbing material

- Dimensions, concentrations, tolerances, and location of neutron-absorbing and moderating materials, including neutron poisons and shielding

- Dimensions and tolerances of any floodable voids, including flux traps, inside the packaging

- Dimensions and tolerances of the overall package that affect the physical separation of the SNF contents in package arrays

- Information on control rod assemblies, shrouds, or other fuel assembly components included with the SNF, as applicable to the criticality evaluation. All information presented in the text, drawings, figures, and tables should be consistent with each other and with that used in the criticality evaluation. The drawings are the authoritative source of dimensions, tolerances, and material composition of components important to criticality safety.

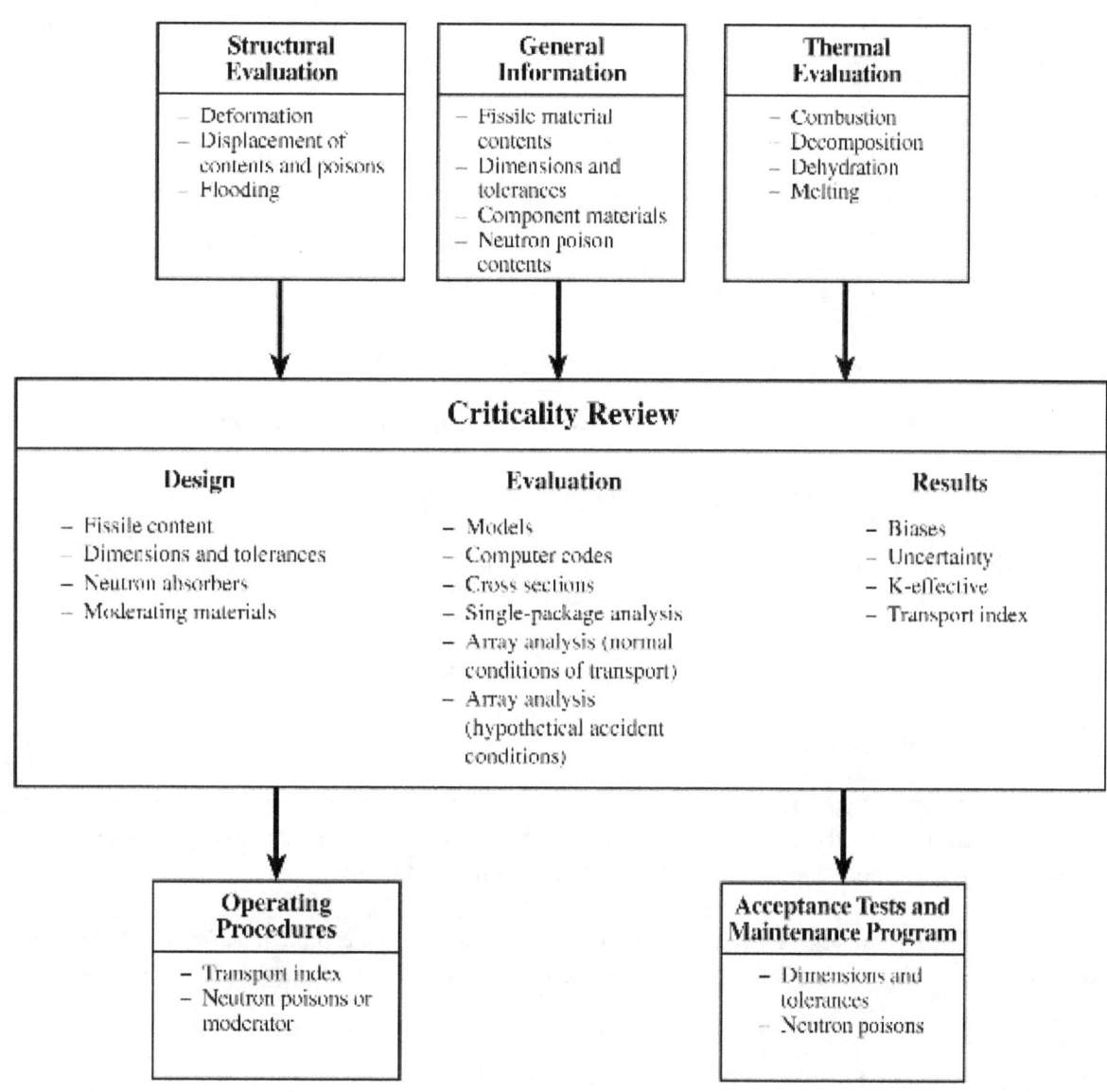

Figure 6-1 SAR Information Flow for the Criticality Review.

6.5.1.2 Codes and Standards

Verify that the established codes and standards used in all aspects of the criticality design and evaluation of the package are identified.

6.5.1.3 Summary Table of Criticality Evaluations

Review the summary table of the criticality evaluation, which should address the following cases, as described in Sections 6.5.4 to 6.5.6 below:

- A single package, under the conditions of 10 CFR 71.55(b), (d), and (e)

- An array of 5N undamaged packages, under the conditions of 10 CFR 71.59(a)(1)

- An array of 2N damaged packages, under the conditions of 10 CFR 71.59(a)(2).

For each case, the table should at least include the maximum value of k_{eff}, the uncertainty, the bias, and the number of packages evaluated in the arrays. The table should also show that the sum of the effective multiplication factor (k_{eff}), two standard deviations (95% confidence), and the bias adjustment does not exceed 0.95 for each case.

Confirm that the summary table illustrates that the package meets the above subcriticality criterion.

6.5.1.4 Transport Index

Based on the number of packages evaluated in the arrays, verify that the SAR determines the appropriate value of N and calculates the criticality transport index correctly. Ensure that this transport index is consistent with that reported in the General Information section of the SAR.

Confirm that the SAR identifies the maximum number of packages that can be transported in the same exclusive-use vehicle. Ensure that this number is clearly distinguished from the value of N used in the criticality evaluation.

6.5.2 Spent Nuclear Fuel Contents

Ensure that the specifications for the SNF used in the criticality evaluation are consistent with those in the General Information section of the SAR. Any differences in the specifications should be clearly identified and justified. Specifications relevant to the criticality evaluation include:

- Types of fuel assemblies or rods (e.g., BWR/PWR) and vendor/model as appropriate

- Dimensions of fuel (including any annular pellets), cladding, fuel-cladding gap, pitch, and rod length

- Number of rods per assembly and locations of guide tubes and burnable poisons (see Section 6.5.3.2)

- Materials and densities

- Active fuel length

- Enrichment (variation by rod if applicable) before irradiation

- Chemical and physical form

- Mass of initial heavy metal per assembly or rod

- Number of fuel assemblies or individual rods per package.

For BWR fuel assemblies, NRC staff does not currently allow any credit for burnup of the fissile material or increase in actinide or fission product poisons during irradiation; therefore, the enrichment should be that of the unirradiated fuel. However, burnup credit is allowed for PWR fuel assemblies. Section 6.4.8 provides guidance on applying burnup credit in the criticality safety analysis of PWR spent fuel. If assemblies contain fuel with several enrichments, the evaluation should either assume the maximum initial enrichment or demonstrate that another approach (e.g., average enrichment) is bounding. Section 6.5.3.2 below discusses consideration of the depletion of burnable poisons.

Determine if the SAR includes any specifications regarding the condition of the SNF. Fuel rods that have been removed from an assembly should be replaced with dummy rods that displace an equal amount of water unless the criticality analyses consider the additional moderation resulting from their absence. (Because of the additional moderation, the contents with less fissile material might be more reactive). These specifications should be included as a condition of approval for the contents in the SER and certificate of compliance.

In general, the package will be designed for numerous types of SNF. The description of the contents should be sufficient to permit a detailed criticality evaluation of each type or to support a conclusion that certain types are bounded by the evaluations performed. The SAR may include separate criticality controls (e.g., number of assemblies, enrichment, transport index) for the various types of SNF evaluated. If the contents include damaged fuel, the maximum extent of damage should be specified and shown to be bounded by the criticality analysis. The review procedures below should address the evaluation for each contents as appropriate.

6.5.3 General Considerations for Criticality Evaluations

The considerations discussed below are applicable to the criticality evaluations of a single package, arrays of packages under normal conditions of transport, and arrays under hypothetical accident conditions.

General guidance for preparing criticality evaluations of transportation packages is provided in NUREG/CR-5661. Guidance for package arrays is provided in NUREG-1646.

6.5.3.1 Model Configuration

Examine the Structural Evaluation and Thermal Evaluation sections of the SAR to determine the effects of the normal conditions of transport and hypothetical accident conditions on the packaging and its contents. Verify that the models used in the criticality calculation are consistent with these effects.

Examine the sketches or figures of the model used for the criticality calculations. Verify that the dimensions and materials are consistent with those in the drawings of the actual package. Differences should be identified and justified. Within the specified tolerance range, dimensions should be selected to result in the highest reactivity.

Verify that the SAR considers deviations from nominal design configurations. For example, the fuel assemblies might not always be centered in each basket compartment, and the basket might not be exactly centered in the package. In addition to a fully flooded package, the SAR should address preferential flooding as appropriate. This includes flooding of the fuel-cladding gap and other regions (e.g., flux traps) for which water density might not be uniform in a flooded package.

Determine whether the SAR includes a heterogeneous model of each fuel rod or homogenizes the entire assembly. With current computational capability, homogenization should generally be avoided. If such homogenization is used, the SAR must demonstrate that it is applied correctly or conservatively. As a minimum, this demonstration should include calculation of the multiplication factor of one assembly and several benchmark experiments (see Section 6.5.7) using both homogeneous and heterogeneous models.

6.5.3.2 Material Properties

Verify that the appropriate mass densities and atom densities are provided for all materials used in the models of the packaging and contents. Material properties should be consistent with the condition of the package under the tests of 10 CFR 71.71 and 10 CFR 71.73, and any differences between normal conditions of transport and hypothetical accident conditions should be addressed. The sources of the data on material properties should be referenced.

No more than 75% of the specified minimum neutron poison concentration of the packaging should generally be considered in the criticality evaluation. In addition, because of differences in net reactivity due to depletion of fissile material and burnable poisons, no credit should be taken for burnable poisons in the fuel. Ensure that neutron absorbers and moderators (e.g., poisons and neutron shielding) are properly controlled during fabrication to meet their specified properties. Such information should be discussed in more detail in the Acceptance Tests and Maintenance Program section of the SAR. Additional guidance on neutron poisons is provided in NUREG-1647.

Review materials to identify any criticality properties that could degrade during the service life of the packaging. If appropriate, ensure that specific controls are in place to assure the effectiveness of the packaging during its service life. Such information should also be discussed in more detail in the Acceptance Tests and Maintenance Program or Operating Procedures sections of the SAR.

6.5.3.3 Computer Codes and Cross Section Libraries

Both Monte Carlo and deterministic computer codes may be used for criticality calculations. Monte Carlo codes are generally better suited to analyzing three-dimensional geometry and, therefore, are more widely used to evaluate SNF cask designs. Verify that the SAR uses an appropriate computer code for the criticality evaluation. Commonly used codes such as SCALE/KENO (NUREG/CR-0200) and MCNP (LA-12625-M) should be clearly referenced. KENO is a multigroup code that is part of the SCALE sequence, while MCNP permits the use of continuous cross sections. Other codes should be described in the SAR, and appropriate supplemental information should be provided.

Ensure that the criticality evaluations use an appropriate cross section library. If multigroup cross sections are used, confirm that the neutron spectrum of the package has been appropriately considered for collapsing the group structure and that the cross sections are properly processed to account for resonance absorption and self-shielding. The use of KENO as part of the SCALE sequence will directly enable such processing. Some cross section sets include data for fissile and fertile nuclides (based on a potential scattering cross section, s_p) that can be input by the user. If the applicant has used a stand-alone version of KENO, ensure that potential scattering has been properly considered. Additional information addressing cross section concerns is provided in an NRC information notice (NRC IN 91-26) and NUREG/CR-6328.

In addition to cross section information, other key input data for the criticality calculations should be identified. These include number of neutrons per generation, number of generations, convergence criteria, mesh selection, etc., depending on the code used. The SAR should also include at least one representative input file for a single package, undamaged array, and damaged array evaluation. Verify that information regarding the model configuration, material properties, and cross sections is properly input into the code.

At least one representative output file (or key sections) should generally also be included in the SAR. Ensure that the calculation has properly converged and that the calculated multiplication factors from the output files agree with those reported in the evaluation.

6.5.3.4 Demonstration of Maximum Reactivity

Verify that the SAR evaluates each type of SNF included as allowable contents or clearly demonstrates that some types are bounded by other evaluations.

Ensure that the analysis determines the optimum combination internal moderation (within the package) and interspersed moderation (between packages), as applicable. Confirm that preferential flooding of different regions within the package is considered as appropriate. As noted in Section 6.5.2, the maximum allowable amount of fissile material may not be the most reactive.

Verify that the analyses demonstrate the most reactive of the three cases listed in Section 6.5.1.3 above (single package, array of undamaged packages, and array of damaged packages) for each of the different types of SNF, as applicable. Assumptions and approximations should be clearly identified and justified.

Additional guidance on determining the most reactive configurations is presented in NUREG/CR-5661.

6.5.3.5 Confirmatory Analyses

The review should include a confirmatory analysis of the criticality calculations reported in the SAR. As a minimum, the reviewer should perform an independent calculation of the most reactive case, as well as sensitivity analyses to confirm that the most reactive case has been correctly identified. In deciding the level of effort necessary to perform independent confirmatory calculations, the reviewer should consider the following three factors: (1) the calculational method (computer code) used by the applicant, (2) the degree of conservatism in the applicant's assumptions and analyses, and (3) how large a margin exists between the calculated result and the acceptance criterion of $k_{eff} \leq 0.95$. As with any design and review, a small margin below the acceptance criterion and/or small degree of conservatism necessitate a more extensive analysis.

The reviewer should generally model the package independently and should use a different code and cross section set from that used in the SAR. If the reported k_{eff} for the worst case is substantially lower than the acceptance criterion of 0.95, a simple model known to produce very conservative results may be all that is necessary for the independent calculations. A review is not expected to validate the applicant's calculations but should assure that the regulations and acceptance criteria are met.

When the value of k_{eff} is highly sensitive to small variations in design features, contents specifications, or the effects of the hypothetical accident conditions, the reviewer should confirm that such variations are appropriately considered.

6.5.4 Single Package Evaluation

6.5.4.1 Configuration

Ensure that the criticality evaluation demonstrates that a single package is subcritical under both normal conditions of transport and hypothetical accident conditions. The evaluations should consider:

- SNF in its most reactive credible configuration consistent with the condition of package and the chemical and physical form of the contents

- Water moderation to the most reactive credible extent, including water inleakage into the containment system as specified in 10 CFR 71.55(b)

- Full water reflection on all sides of the package, including close reflection of the containment system or reflection by the package materials, whichever is more reactive, as specified in 10 CFR 71.55(b)(3).

6.5.4.2 Results

Confirm that the results of the criticality calculations are consistent with the information presented in the summary table discussed in Section 6.5.1.3.

Verify also that the package meets the additional specifications of 10 CFR 71.55(d)(2) through 10 CFR 71.55(d)(4) under normal conditions of transport. These requirements address subcriticality, alteration of the geometric form of the contents, inleakage of water, and effectiveness of the packaging.

6.5.5 Evaluation of Package Arrays under Normal Conditions of Transport

6.5.5.1 Configuration

Ensure that the criticality evaluation demonstrates that an array of 5N packages is subcritical under normal conditions of transport. The evaluation should consider:

- The most reactive configuration of the array, e.g., pitch, package orientation, etc., with nothing (including moderator) between the packages

- The most reactive credible configuration of the packaging and its contents. (Because water does not leak into a spent-fuel package *under normal conditions of transport*, water inleakage need not be assumed.)

- Full water reflection on all sides of the array (unless the array is infinite).

6.5.5.2 Results

Verify that the most reactive array conditions are clearly identified and that the results of the analysis are consistent with the information presented in the summary table discussed in Section 6.5.1.3 above.

Confirm that the appropriate N value is used in determination of the transport index. The appropriate N should be the smaller value which assures subcriticality for 5N packages under normal conditions of transport or 2N packages under hypothetical accident conditions, as discussed in the next section.

6.5.6 Evaluation of Package Arrays under Hypothetical Accident Conditions

6.5.6.1 Configuration

Ensure that the criticality evaluation demonstrates that an array of 2N packages is subcritical under hypothetical accident conditions. The evaluation should consider:

- The most reactive configuration of the array, e.g., pitch, package orientation, etc.

- Optimum interspersed hydrogenous moderation (between packages)

- The most reactive credible configuration of the packaging and its contents, including inleakage of water and internal moderation

- Full water reflection on all sides of the array (unless the array is infinite).

6.5.6.2 Results

Verify that the most reactive array conditions are clearly identified and that the results of the analysis are consistent with the information presented in the summary table discussed in Section 6.5.1.3 above.

Confirm that the appropriate N value is used in determining the transport index. The appropriate N should be the smaller value which assures subcriticality for 2N packages under hypothetical accident conditions or 5N packages under normal conditions of transport, as discussed in the previous section.

6.5.7 Benchmark Evaluations

Ensure that the computer codes for criticality calculations are benchmarked against critical experiments. Verify that the analysis of the benchmark experiments used the same computer code, hardware, and cross section library as those used to calculate the multiplication factor for the package evaluations. The calculated k_{eff} of the cask should then be adjusted to include the appropriate biases and uncertainties from the benchmark calculations.

Additional information on benchmarking criticality evaluations for SNF is provided in NUREG/CR-6361.

6.5.7.1 Experiments and Applicability

Review the general description of the benchmark experiments and confirm that they are appropriately referenced.

The applicant should justify and the reviewer should verify that the benchmark experiments are applicable to the actual package design. The benchmark experiments should have, to the maximum extent possible, the same materials, neutron spectrum, and configuration as the package evaluations. Key package parameters that should be compared with those of the benchmark experiments include type of fissile material, enrichment, H/U ratio (dependent largely on rod pitch and diameter), poisoning, reflector material, and configuration. Confirm that differences between the package and benchmarks are discussed and properly considered.

In addition, the SAR should address the overall quality of the benchmark experiments and the uncertainties in experimental data (e.g., mass, density, dimensions, etc.). Ensure that these uncertainties are treated in a conservative manner, i.e., they result in a lower calculated multiplication factor for the benchmark experiment.

6.5.7.2 Bias Determination

Examine the results of the calculations for the benchmark experiments and the method used to account for biases, including the contribution from uncertainties in experimental data.

Assess that a sufficient number of appropriate benchmark experiments are analyzed and that the results of these benchmark calculations are used to determine an appropriate bias for the package calculations. The applicant should check benchmark comparisons for trends in the bias with respect to parameter variations (such as pitch-to-rod-diameter ratio, assembly separation, reflector material, neutron absorber material, etc.). Verify that only negative biases are considered, with positive bias results (values which decrease k_{eff} when applied) treated as zero bias.

Statistical and convergence uncertainties of both benchmark and package calculations should also be addressed. The uncertainties should be applied to at least the 95-percent confidence level. As a general rule, if the acceptability of the result depends on these rather small differences, reviewers should question the overall degree of conservatism of the calculations. Considering the current availability of computer resources, a sufficient number of neutron histories can readily be used so that the treatment of these uncertainties should not significantly affect the results.

6.5.8 Burnup Credit

Review the burnup credit analysis to determine compliance with the staff guidance outlined in Sections 6.4.8.1 through 6.4.8.6. The guidance provides a design-specific basis for granting burnup credit, based on actinide composition.

The staff's guidance for burnup credit considerations are based on investigations which have been performed, both within the United States and by other countries, in an effort to understand and document the related phenomena. The staff will issue additional guidance, as necessary, when more information is obtained from its research program on burnup credit and as experience is gained through future licensing activities.

6.5.9 Appendix

The appendix may include a list of references, copies of applicable references if not generally available to the reviewer, computer code descriptions, input and output files, test results, and any other appropriate supplemental information.

6.6 EVALUATION FINDINGS

The criticality review should result in the following findings, as appropriate:

6.6.1 Description of Criticality Design

The staff has reviewed the description of the packaging design and concludes that it provides an adequate basis for the criticality evaluation.

The staff has reviewed the summary information of the criticality design and concludes that it indicates the package is in compliance with the requirements of 10 CFR Part 71.

6.6.2 Spent Nuclear Fuel Contents

The staff has reviewed the description of the SNF contents and concludes that it provides an adequate basis for the criticality evaluation.

6.6.3 General Considerations for Criticality Evaluations

The staff has reviewed the criticality description and evaluation of the package and concludes that it addresses the criticality safety requirements of 10 CFR Part 71.

6.6.4 Single Package Evaluation

The staff has reviewed the criticality evaluation of a single package and concludes that it is subcritical under the most reactive credible conditions.

6.6.5 Evaluation of Package Arrays under Normal Conditions of Transport

The staff has reviewed the criticality evaluation of the most reactive array of 5N packages and concludes that it is subcritical under normal conditions of transport.

6.6.6 Evaluation of Package Arrays under Hypothetical Accident Conditions

The staff has reviewed the criticality evaluation of the most reactive array of 2N packages and concludes that it is subcritical under hypothetical accident conditions.

6.6.7 Benchmark Evaluations

The staff has reviewed the benchmark evaluation of the calculations and concludes that they are sufficient to determine an appropriate bias and uncertainties for the criticality evaluation of the package.

6.6.8 Burnup Credit

The staff has reviewed the criticality evaluation for granting burnup credit and concludes that the associated fuel loading curve is appropriate.

6.7 REFERENCES

LA-12625-M   Los Alamos National Laboratory, "MCNP 4A, Monte Carlo N-Particle Transport Code System," LA-12625-M, Los Alamos, NM, December 1993.

NRC IN 91-26

U.S. Nuclear Regulatory Commission, "Potential Nonconservative Errors in the Working Format Hansen-Roach Cross-Section Set Provided with the KENO and SCALE Codes," Information Notice 91-26, U.S. Government Printing Office, Washington, D.C., April 15, 1991.

NUREG-1646

U.S. Nuclear Regulatory Commission, "Criticality Analysis of Transportation-Package Arrays," NUREG-1646, U.S. Government Printing Office, Washington, D.C., January 1999.

NUREG-1647

U.S. Nuclear Regulatory Commission, "Use of Neutron Poisons for Criticality Control in Transportation Packages," NUREG-1647, U.S. Government Printing Office, Washington, D.C., January 1999.

NUREG/CR-0200

U.S. Nuclear Regulatory Commission, "SCALE: A Modular Code System for Performing Standardized Computer Analyses for Licensing Evaluation," NUREG/CR-0200, Vol. 2, Part 2, Rev. 4, U.S. Government Printing Office, Washington, D.C., April 1995.

NUREG/CR-5661

U.S. Nuclear Regulatory Commission, "Recommendations for Preparing the Criticality Safety Evaluation of Transportation Packages," NUREG/CR-5661, U.S. Government Printing Office, Washington, D.C., April 1997.

NUREG/CR-6328

U.S. Nuclear Regulatory Commission, "Adequacy of the 123-Group Cross-Section Library for Criticality Analyses of Water-Moderated Uranium Systems," NUREG/CR-6328, U.S. Government Printing Office, Washington, D.C., August 1995.

NUREG/CR-6361

U.S. Nuclear Regulatory Commission, "Criticality Benchmark Guide for Light-Water-Reactor Fuel in Transportation and Storage Packages," NUREG/CR-6361, U.S. Government Printing Office, Washington, D.C., March 1997.

7 OPERATING PROCEDURES REVIEW

7.1 REVIEW OBJECTIVE

The objective of this review is to verify that the operating procedures comply with the requirements of 10 CFR 71 and ensure that the package will be operated in a manner consistent with the conditions assumed in its evaluation for approval.

7.2 AREAS OF REVIEW

The SAR should be reviewed for adequacy of the operating procedures description. Areas of review include the following:

7.2.1 Package Loading
7.2.1.1 Preparation for Loading
7.2.1.2 Loading of Contents
7.2.1.3 Preparation for Transport

7.2.2 Package Unloading
7.2.2.1 Receipt of Package from Carrier
7.2.2.2 Preparation for Unloading
7.2.2.3 Removal of Contents

7.2.3 Preparation of Empty Package for Transport

7.2.4 Other Procedures

7.2.5 Appendix

7.3 REGULATORY REQUIREMENTS

Regulatory requirements of 10 CFR Part 71 applicable to package operations and the operating procedures review are as follows:

7.3.1 Package Loading

The SAR must identify established codes and standards applicable to the operating procedures. [10 CFR 71.31(c)]

The SAR for a fissile material shipment must include any proposed special controls and precautions for transport, loading, unloading, and handling and any proposed special controls in case of accident or delay. [10 CFR 71.35(c)]

Packages must be prepared for transport so that in still air at 38°C (100°F) and in the shade, no accessible surface of a package would have a temperature exceeding 85°C (185°F) in an exclusive-use shipment. [10 CFR 71.43(g)] (Temperature limits for non exclusive-use shipments are assumed not to apply to spent nuclear fuel (SNF) packages.)

Packages which require exclusive use shipment because of external radiation levels must be controlled by providing written instructions to the carrier. [10 CFR 71.47(b), 10 CFR 71.47(c), and 10 CFR 71.47(d)]

Before each shipment, the package must be verified to be proper for the contents to be shipped. [10 CFR 71.87(a)]

Before each shipment, the package must be verified to be in unimpaired physical condition. [10 CFR 71.87(b)]

Before each shipment, each closure device of the package, including any specified gasket, must be verified to be properly installed and secured and free of defects. [10 CFR 71.87(c)]

Before each shipment, any system for containing liquid must be verified to be adequately sealed and to have adequate space or other specified provision for expansion of the liquid. [10 CFR 71.87(d)]

Before each shipment of licensed material any pressure relief device must be verified to be operable and properly set. [10 CFR 71.87(e)]

Before each shipment, it must be determined that the package has been loaded and closed appropriately. [10 CFR 71.87(f)]

Before each shipment of fissile material, it must be determined that any moderator or neutron absorber, if specified, is present and in proper condition. [10 CFR 71.87(g)]

Before each shipment, any structural part of the package that could be used to lift or to tie-down the package during transport must be rendered inoperable for that purpose unless it satisfies the design requirements of 10 CFR 71.45. [10 CFR 71.87(h)]

Before each shipment, the level of non-fixed (removable) radioactive contamination on the external surfaces of each package offered for shipment must be as low as is reasonably achievable (ALARA), and within the limits specified in DOT regulation 49 CFR 173.443. [10 CFR 71.87(i)]

External radiation levels around the package and around the vehicle, if applicable, will not exceed the limits specified in 10 CFR 71.47 at any time during transportation. [10 CFR 71.87(j)]

Accessible package surface temperatures will not exceed the limits specified in 10 CFR 71.43(g) at any time during transportation. [10 CFR 71.87(k)]

Before delivery of a package to a carrier for transport, the licensee must send or make available any special instructions needed to safely open the package to the consignee for the consignee's use in accordance with 10 CFR 20.1906 (e). [10 CFR 71.89]

7.3.2 Package Unloading

The application for a fissile material shipment must include provisions for complying with 10 CFR 20.1906 and any proposed special controls and precautions for unloading and handling. [10 CFR 71.35(c) and 10 CFR 71.89]

7.3.3 Preparation of Empty Package for Transport

Before each shipment, perform the necessary inspections and tests to ensure that the level of non-fixed (removable) radioactive contamination on the external surfaces of each package offered for shipment is ALARA, and within the limits specified in DOT regulation 49 CFR 173.443. [10 CFR 71.87(i)]

7.3.4 Other Procedures

The application for a fissile material shipment must include any proposed special controls and precautions for transport, loading, unloading, and handling and any proposed special controls in case of accident or delay. [10 CFR 71.35(c)]

7.4 ACCEPTANCE CRITERIA

The operating procedures should be presented and discussed sequentially in the actual order of performance.

7.4.1 Package Loading

In addition to the regulatory requirements identified in Section 7.3.1, leakage testing of the package should meet the assembly verification leakage test requirements specified in ANSI N14.5.

7.4.2 Package Unloading

The regulatory requirements in Section 7.3.2 identify the acceptance criteria.

7.4.3 Preparation of Empty Package for Transport

In addition to the regulatory requirements identified in Section 7.3.3, the interior of the packaging should be properly decontaminated and closed in accordance with the requirements of 49 CFR 173.428.

7.4.4 Other Procedures

In addition to the regulatory requirements identified in Section 7.3.4, the package should be properly closed and delivered to the carrier in such a condition that subsequent transport will not reduce the effectiveness of the packaging.

7.5 REVIEW PROCEDURES

The following procedures are generally applicable to the operating procedures review of all SNF transportation packages. Since packages for shipment of SNF are generally intended to be shipped by exclusive-use, only exclusive-use shipments are assumed in the following SRP review procedures.

The operating procedures review is based in part on the descriptions and evaluations presented in the General Information, Structural Evaluation, Thermal Evaluation, Containment Evaluation, Shielding Evaluation, and Criticality Evaluation sections of the SAR. Examples of SAR information flow into and within the operating procedures review are shown in Figure 7-1.

Figure 7-1 SAR Information Flow for the Operating Procedures Review.

The operating procedures presented in the SAR should not be expected to be detailed procedures that could be implemented without expansion. Rather, the operating procedures should be an outline that focuses upon those steps that are important to assuring that the package is operated in a manner consistent with its evaluation for approval. Detailed procedures not important to safety should not be included in the SAR. Procedural steps should normally be presented in sequential order, as applicable. Information on both the detailed procedures and the brief procedures included with an application can be found in NUREG/CR-4775.

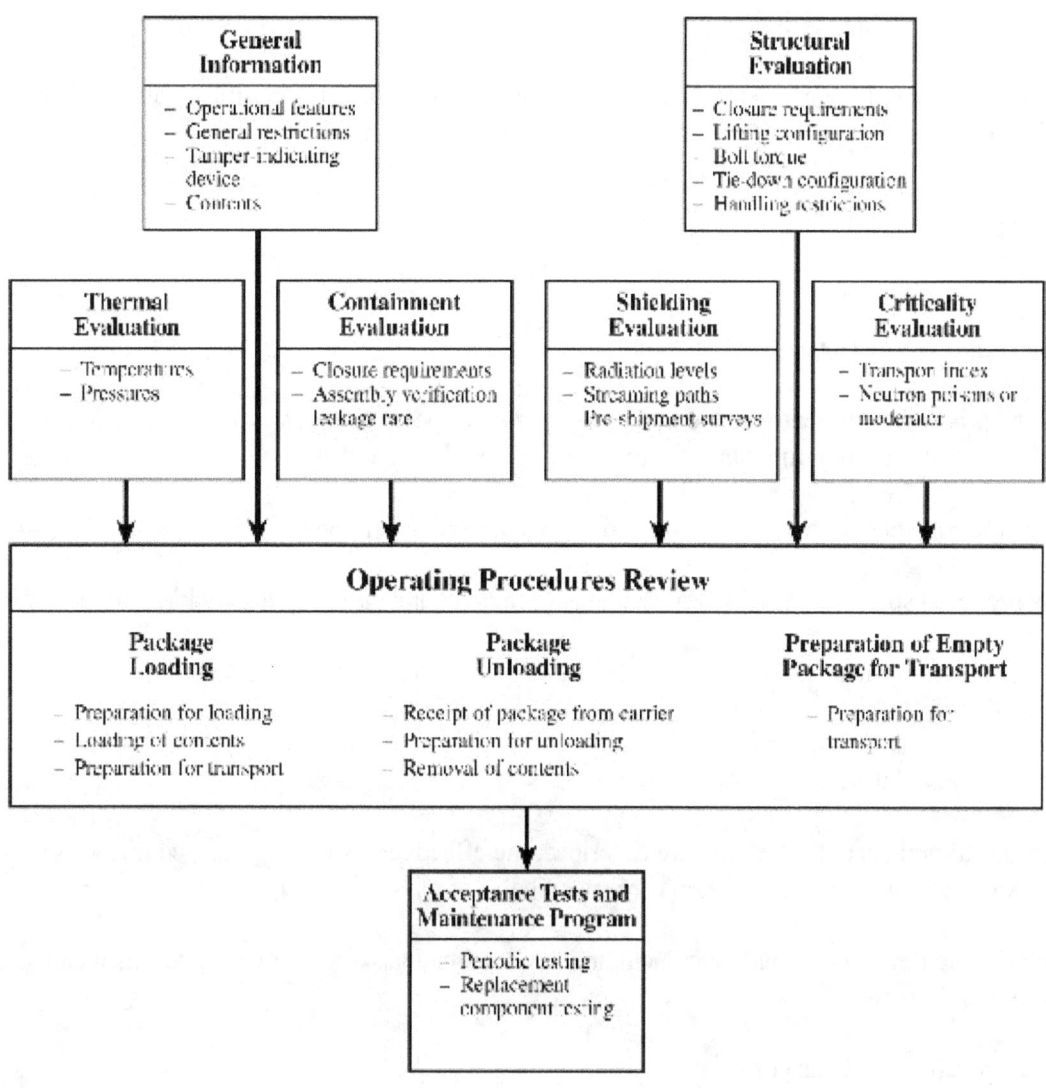

7.5.1 Package Loading

7.5.1.1 Preparation for Loading

Review the procedures presented sequentially in the order of performance for loading and preparing the package for transport. At a minimum, the procedures should ensure that:

- The package is loaded and closed in accordance with written instructions

- The contents are authorized by the certificate of compliance, including the use of a secondary container or containment as appropriate

- The use of the package complies with the conditions of approval in the certificate of compliance, including verification that required maintenance has been performed

- The package is in unimpaired physical condition

- Any proposed special controls and precautions for handling the package are provided.

7.5.1.2 Loading of Contents

Review the procedures presented sequentially in the order of performance for loading the package contents for transport. At a minimum, the procedures should ensure that:

- Special handling equipment needed for loading and unloading is provided

- Any proposed special controls and precautions for loading and handling the package are provided

- Any moderator or neutron absorber, if specified, is present and in proper condition

- The package has been loaded and closed appropriately in accordance with the specified bolt torques and bolt-tightening sequences

- Methods to drain and dry the cask are described, the effectiveness of the proposed methods is discussed, and vacuum drying criteria are specified

- Each closure device of the package, including any specified gaskets, is properly installed and secured and free of defects.

7.5.1.3 Preparation for Transport

Review the procedures presented sequentially in the order of performance for preparing the package for transport. At a minimum, the procedures should ensure that:

- The level of non-fixed (removable) radioactive contamination on the external surfaces of each package offered for shipment is ALARA, and within the limits specified in DOT regulation 49 CFR 173.443

- Radiation survey requirements of the package exterior are described to ensure that limits specified in 10 CFR 71.47 are met

- The temperature survey requirements of the package exterior ensure that limits specified in 10 CFR 71.43(g) are implemented

- Leakage testing of the package meets the assembly verification leakage test requirements specified in ANSI N14.5

- A tamper indicating device is incorporated which, while intact, indicates that the package has not been opened by unauthorized persons

- Any system for containing liquid is adequately sealed and has adequate space or other specified provision for expansion of the liquid

- A check is made to ensure that any pressure relief device is operable and properly set

- Any structural part of the package that could be used to lift or to tie-down the package during transport is rendered inoperable for that purpose unless it satisfies the design requirements of 10 CFR 71.45

- Any proposed special controls and precautions for transport, handling and any proposed special controls in case of accident or delay are specified

- Written instructions to the carrier are provided for packages which require exclusive use shipment because of external radiation levels [10 CFR 71.47(b), 10 CFR 71.47(c), and 10 CFR 71.47(d)]

- Before delivery of a package to a carrier for transport, the licensee has sent or made available to the consignee any special instructions needed to safely open the package, in accordance with 10 CFR 20.1906(e).

7.5.2 Package Unloading

In general, the unloading procedures are the reverse of the loading procedures. If applicable, procedures for special controls and precautions to ensure safe removal of fission gases, contaminated coolant, and solid contaminants should be presented and discussed.

7.5.2.1 Receipt of Package from Carrier

Review the procedures presented sequentially in the order of performance for receiving the package from the carrier. At a minimum, the procedures should ensure that:

- The requirements of 10 CFR 20.1906 are met

- The package is examined for visible external damage

- Steps to define actions to be taken when the tamper indicating device is not intact, or surface contamination or radiation survey levels are too high are provided

- A list of any special handling equipment needed for unloading and handling the package is provided

- Any proposed special controls and precautions for unloading and handling the package are provided.

7.5.2.2 Preparation for Unloading

Review the procedures presented sequentially in the order of performance for preparing the package for unloading following receipt. At a minimum, the procedures should ensure that:

- Procedures controlling the radiation level limits on unloading operations are provided

- Procedures for the safe removal of, if any, fission gases, contaminated coolants, and solid contaminants are provided.

7.5.2.3 Removal of Contents

Review the procedures presented sequentially in the order of performance for removing the contents following package receipt. At a minimum, the procedures should ensure that:

- The closure is removed appropriately

- The contents are removed appropriately

- A verification is made that the contents are completely removed.

7.5.3 Preparation of Empty Package for Transport

Review the procedures presented sequentially in the order of performance for preparing an empty package for transport. At a minimum, the procedures should ensure that:

- The packaging is empty

- Appropriate inspections and tests of the package are performed before transport, to ensure that the requirements of 10 CFR 71.87(i) are met

- Special preparations of the packaging, to ensure that the interior of the packaging is properly decontaminated and closed in accordance with the requirements of 49 CFR 173.428, are described.

7.5.4 Other Procedures

Other procedures, as appropriate, should be included.

7.5.5 Appendix

The appendix may include a list of references, copies of any applicable references not generally available to the reviewer, test results, and any other appropriate supplemental information.

7.6 EVALUATION FINDINGS

The operating procedures review should conclude that the applicant's SAR presents acceptable operating sequences, guidance, and generic procedures for key operations and result in the following findings, as appropriate:

7.6.1 Package Loading

The staff has reviewed the proposed special controls and precautions for transport, loading, and handling and any proposed special controls in case of accident or delay, and concludes that they satisfy 10 CFR 71.35(c).

The staff has reviewed the description of the radiation survey requirements of the package exterior and concludes that the limits specified in 10 CFR 71.47 will be met.

The staff has reviewed the description of the temperature survey requirements of the package exterior and concludes that the limits specified in 10 CFR 71.43(g) will be met.

The staff has reviewed the description of the routine determinations for package use prior to transport, and concludes that the requirements of 10 CFR 71.87 will be met.

The staff has reviewed the description of the special instructions (if applicable) needed to safely open a package and concludes that the procedures for providing the special instruction to the consignee are in accordance with the requirements of 10 CFR 71.89.

7.6.2 Package Unloading

The staff has reviewed the proposed special controls and precautions for unloading and handling and concludes that they satisfy 10 CFR 71.35(c).

7.6.3 Preparation of Empty Package for Transport

The staff has reviewed the description of the routine determinations for package use prior to transport, and concludes that the requirements of 10 CFR 71.87 will be met.

7.6.4 Other Procedures

The staff has reviewed all other applicable proposed special controls and precautions for transport, loading, unloading, and handling and concludes that they satisfy 10 CFR 71.35(c).

7.7 REFERENCES

ANSI N14.5   Institute for Nuclear Materials Management, ANSI N14.5, "American National Standard for Leakage Tests on Packages for Shipment of Radioactive Materials," New York, NY, 1987.

NUREG/CR-4775   U.S. Nuclear Regulatory Commission, "Guide for Preparing Operating Procedures for Shipping Packages," NUREG/CR-4775, U.S. Government Printing Office, Washington, D.C., December 1988.

8 ACCEPTANCE TESTS AND MAINTENANCE PROGRAM REVIEW

8.1 REVIEW OBJECTIVE

The objectives of this review are to verify that the acceptance tests for the packaging comply with the requirements of 10 CFR Part 71 for the package design and that a maintenance program will ensure acceptable packaging performance throughout its service life.

8.2 ACCEPTANCE TESTS

8.2.1 Areas of Review

The SAR should be reviewed for adequacy of the description of the acceptance tests to be performed on the packaging. Areas of review include the following:

8.2.1.1 Visual Inspections and Measurements
8.2.1.2 Weld Inspections
8.2.1.3 Structural and Pressure Tests
8.2.1.4 Leakage Tests
8.2.1.5 Component Tests
8.2.1.6 Shielding Tests
8.2.1.7 Neutron Absorber Tests
8.2.1.8 Thermal Tests
8.2.1.9 Appendix

8.2.2 Regulatory Requirements

Regulatory requirements of 10 CFR Part 71 applicable to the acceptance tests review are as follows:

The SAR should identify established codes, standards, and specific provisions of the quality assurance program that are applicable to the acceptance tests to be performed on the packaging. [10 CFR 71.31(c) and 10 CFR 71.37(b)]

Before first use, the fabrication of each packaging must be verified to be in accordance with the approved design. [10 CFR 71.85(c)]

Before first use, each packaging must be inspected for cracks, pinholes, uncontrolled voids, or other defects that could significantly reduce its effectiveness. [10 CFR 71.85(a)]

Before first use, if the maximum normal operating pressure (MNOP) of a package exceeds 35 kPa (5 lbf/in^2) gauge, the containment system of each packaging must be tested at an internal pressure at least 50 percent higher than MNOP to verify its capability to maintain structural integrity at that pressure. [10 CFR 71.85(b)]

Before first use, if applicable, the amount and the distribution of the neutron absorbing materials or moderators must be verified to meet the design specification. [10 CFR 71.87(g)]

Before first use, each packaging must be conspicuously and durably marked with its model number, serial number, gross weight, and a package identification number assigned by NRC. [10 CFR 71.85(c)]

The licensee must perform any tests deemed appropriate by NRC. [10 CFR 71.93(b)]

8.2.3 Acceptance Criteria

In addition to the regulatory requirements identified in Section 8.2.2, the SAR should discuss the package tests to be performed and the acceptance criteria to demonstrate structural, leakage, shielding, and heat transfer performance. Fabrication, welding, and examination of components are acceptable when performed in accordance with the recommended sections and subsections of the ASME Boiler and Pressure Vessel (B&PV) Code given in Section 1.5.2.6, Table 1-1 and Table 1-2 of this SRP. Leakage testing of the packaging should be accomplished in accordance with ANSI N14.5. Fabrication, examination, and acceptance testing of lifting trunnions should be conducted in accordance with NUREG-0612, ANSI N14.6, or other appropriate specification.

8.2.4 Review Procedures

The following procedures are generally applicable to the acceptance tests review of all spent nuclear fuel (SNF) packages.

The acceptance tests review is based in part on the descriptions and evaluations presented in the General Information, Structural Evaluation, Thermal Evaluation, Containment Evaluation, Shielding Evaluation, Criticality Evaluation, and Operating Procedures sections of the SAR and follows the sequence established to evaluate the packaging against applicable 10 CFR Part 71 requirements. Examples of SAR information flow into and within the acceptance tests review are shown in Figure 8-1.

The commitments specified in the Acceptance Tests and Maintenance Program section of the SAR are often incorporated by reference into the certificate of compliance as conditions of package approval.

Verify that the following tests, as applicable, are performed prior to the first use of the packaging. Information presented on each test should include, as a minimum, a description of the test, the test procedure, and the acceptance criteria. Confirm that the established codes, standards, and specific provisions of the quality assurance program used in all aspects of the testing of the packaging are identified.

Additional guidance on acceptance tests is provided in NUREG/CR-3854.

8.2.4.1 Visual Inspections and Measurements

Ensure that visual inspections are performed to verify that the packaging has been fabricated and assembled in accordance with drawings and other requirements specified in the SAR. Dimensions and tolerances specified on the drawings should be confirmed by measurement.

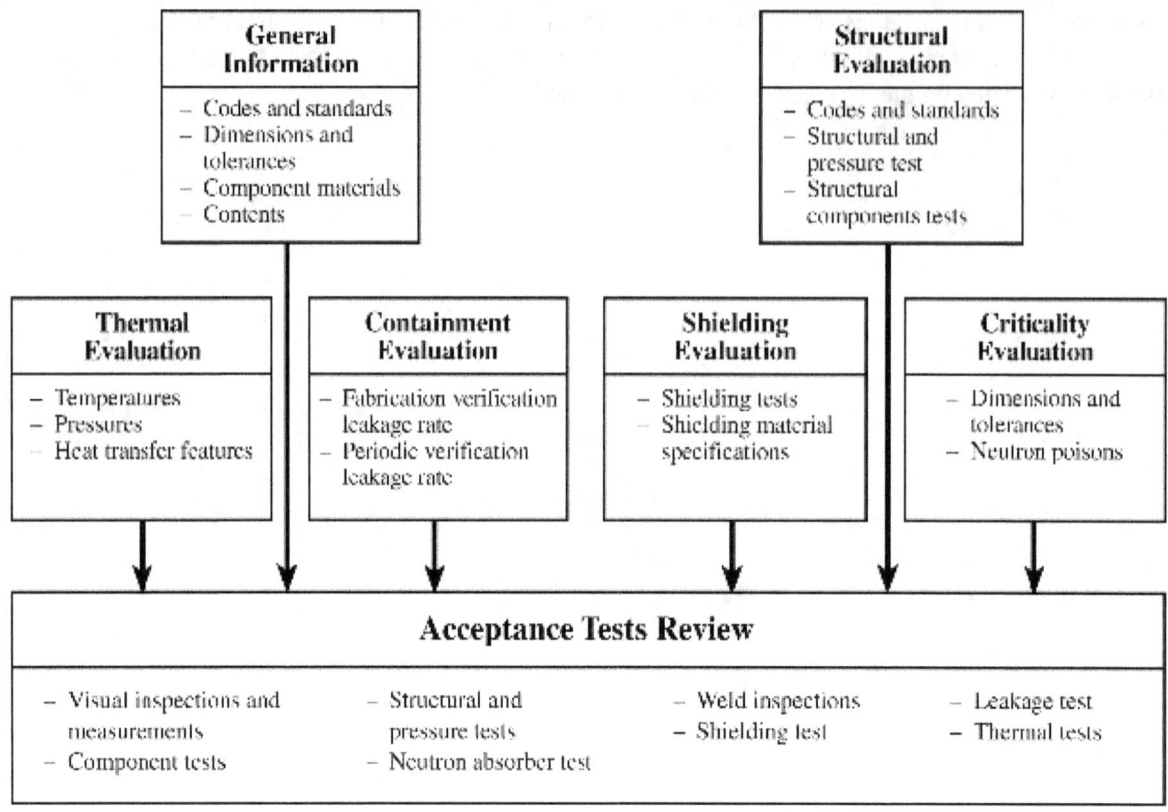

Figure 8-1 SAR Information Flow for the Acceptance Tests Review.

8.2.4.2 Weld Inspections

Verify that weld inspections are performed to verify fabrication in accordance with the drawings, codes, and standards specified in the SAR to control weld quality. Location, type, and size of the welds should be confirmed by measurement. Other specifications for weld performance, inspection, and acceptance should be verified as appropriate.

Additional guidance on welding criteria is provided in NUREG/CR-3019.

8.2.4.3 Structural and Pressure Tests

Verify that the structural or pressure tests are identified and described. Such tests should comply with 10 CFR 71.85(b), as well as applicable codes or standards specified in the SAR. Structural testing of lifting trunnions should be conducted in accordance with NUREG-0612, ANSI N14.6, or other appropriate specification.

8.2.4.4 Leakage Tests

Verify that the containment system of the packaging is subjected to the fabrication leakage tests specified in ANSI N14.5. The acceptable leakage criterion should be consistent with that identified in the Containment Evaluation section of the SAR.

8.2.4.5 Component Tests

Confirm that tests and acceptance criteria for other components are specified as appropriate. Such components include valves, rupture disks, seals, etc.

8.2.4.6 Shielding Tests

Ensure that shielding tests are specified for gamma and neutron radiation, as appropriate. The tests and acceptance criteria should be sufficient to assure no voids or streaming paths exist in the shielding.

8.2.4.7 Neutron Absorber Tests

Verify that appropriate tests are specified to verify the amount and distribution meeting the minimum specification of neutron absorbing material described in the SAR.

8.2.4.8 Thermal Tests

Verify that appropriate tests are specified to demonstrate the heat transfer capability of the packaging. These tests should confirm the heat transfer properties predicted in the Thermal Evaluation section of the SAR.

8.2.4.9 Appendix

The appendix may include a list of references, copies of any applicable references not generally available to the reviewer, and any other appropriate supplemental information.

8.2.5 Evaluation Findings

The acceptance tests review should result in the following findings, as appropriate:

The staff has reviewed the identification of the codes, standards, and provisions of the quality assurance program applicable to the package design and concludes that the requirements specified in 10 CFR 71.31(c) and 10 CFR 71.37 (b) will be met.

The staff has reviewed the description of the preliminary determinations for the package prior to first use and concludes that the requirements of 10 CFR 71.85 and 10 CFR 71.87(g) will be met.

8.3 MAINTENANCE PROGRAM

8.3.1 Areas of Review

The SAR should be reviewed for adequacy of the description of the maintenance program to be performed on the packaging. Areas of review include the following:

8.3.1.1 Structural and Pressure Tests
8.3.1.2 Leakage Tests
8.3.1.3 Component Tests
8.3.1.4 Neutron Absorber Tests
8.3.1.5 Thermal Tests
8.3.1.6 Appendix

8.3.2 Regulatory Requirements

Regulatory requirements of 10 CFR Part 71 applicable to the maintenance program review are as follows:

The SAR should identify established codes, standards, and specific provisions of the quality assurance program that are applicable to the proper maintenance of the packaging. [10 CFR 71.31(c) and 10 CFR 71.37(b)]

The maintenance program should ensure that the packaging is maintained in unimpaired physical condition except for superficial defects such as marks or dents. [10 CFR 71.87(b)]

The presence of a moderator or neutron absorber should be verified to be in proper condition prior to each shipment. [10 CFR 71.87(g)]

The licensee must perform any tests deemed appropriate by NRC. [10 CFR 71.93(b)]

8.3.3 Acceptance Criteria

In addition to the regulatory requirements identified in Section 8.3.2, the maintenance program should include periodic testing requirements, inspections, and replacement criteria and schedules for replacements and repairs of components on an as-needed basis.

8.3.4 Review Procedures

The following procedures are generally applicable to the maintenance program review of all SNF packages.

The maintenance program review is based in part on the descriptions and evaluations presented in the General Information, Structural Evaluation, Thermal Evaluation, Containment Evaluation, Shielding Evaluation, Criticality Evaluation, and Operating Procedures sections of the SAR and follows the sequence established to evaluate the packaging against applicable 10 CFR Part 71 requirements.

Examples of SAR information flow into and within the maintenance program review are shown in Figure 8-2.

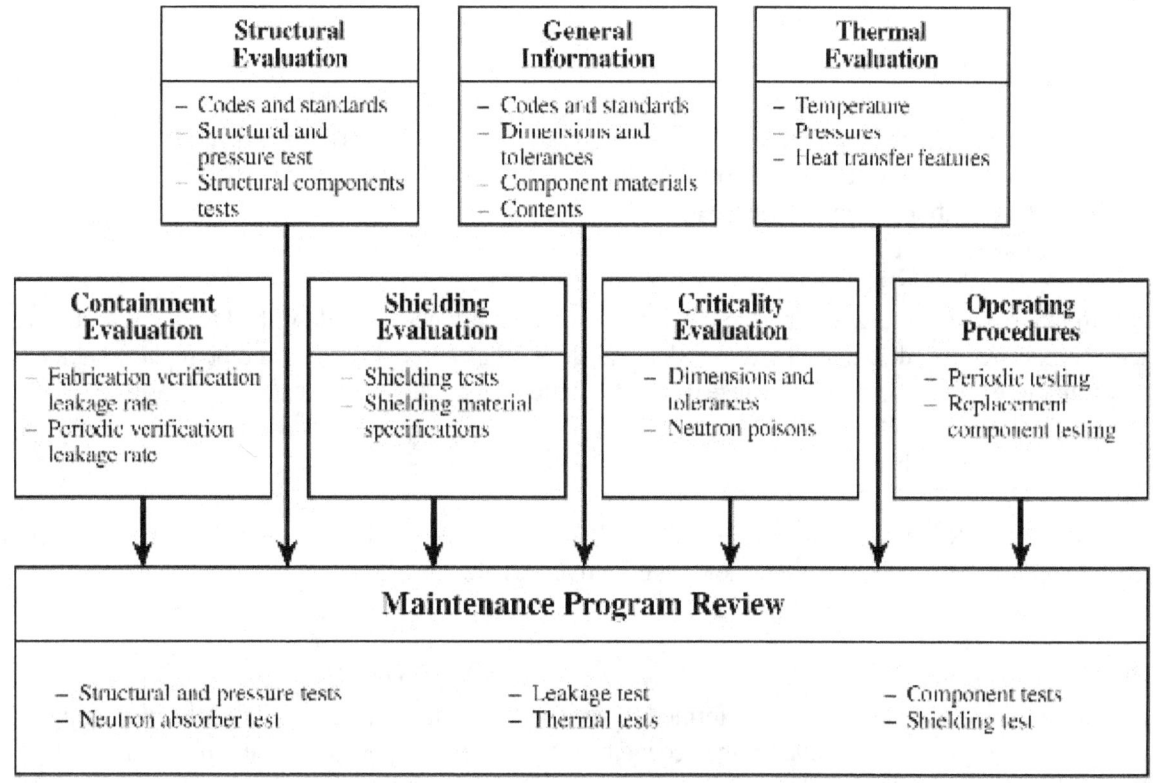

Figure 8-2 SAR Information Flow for the Maintenance Program Review.

The commitments specified in the Acceptance Tests and Maintenance Program section of the SAR are often incorporated by reference into the certificate of compliance as conditions of package approval.

The maintenance program should be adequate to assure that packaging effectiveness is maintained throughout its service life. Verify that the following maintenance tests and inspections are described with schedules and criteria for minor refurbishment and replacement of parts, as applicable. Confirm that the established codes, standards, and specific provisions of the quality assurance program used in all aspects of the maintenance of the packaging are identified.

8.3.4.1 Structural and Pressure Tests

Verify that any structural or pressure tests are identified and described. Such tests would generally be applicable to codes, standards, or other procedures specified in the SAR. Structural testing of lifting trunnions should be conducted in accordance with NUREG-0612, ANSI N14.6, or other appropriate specification.

8.3.4.2 Leakage Tests

Verify that the containment system of the packaging is subjected to the periodic leakage tests specified in ANSI N14.5. The acceptable leakage criterion should be consistent with that identified in the Containment Evaluation section of the SAR.

8.3.4.3 Component Tests

Confirm that periodic tests and replacement schedules for components are described as appropriate. Such components include valves, rupture disks, and seals. Elastomeric seals should be replaced at an interval not to exceed one year. Metal seals should be replaced after each use.

8.3.4.4 Neutron Absorber Tests

Verify that the SAR identifies any process that could result in deterioration of neutron absorbing material and that appropriate tests to ensure packaging effectiveness are specified.

8.3.4.5 Thermal Tests

Appropriate periodic tests should be performed to verify the heat transfer capability of the packaging during its service life. Tests similar to the acceptance tests may be applicable. The typical interval for periodic thermal tests is five years.

8.3.4.6 Appendix

The appendix may include a list of references, copies of any applicable references not generally available to the reviewer, and any other appropriate supplemental information.

8.3.5 Evaluation Findings

The maintenance program review should result in the following findings, as appropriate:

The staff has reviewed the identification of the codes, standards, and provisions of the quality assurance program applicable to maintenance of the packaging and concludes that the requirements specified in 10 CFR 71.31(c) and 10 CFR 71.37 (b) will be met.

The staff has reviewed the description of the routine determinations for package use prior to transport and concludes that the requirements of 10 CFR 71.87(b) and 10 CFR 71.87(g) will be met.

8.4 REFERENCES

ANSI N14.5	Institute for Nuclear Materials Management, ANSI N14.5, "Leakage Tests on Packages for Shipment of Radioactive Materials," New York, NY, 1987.
ANSI N14.6	Institute for Nuclear Materials Management, ANSI N14.6, "Special Lifting Devices for Shipping Containers Weighing 10,000 Pounds (45000 kg) or More for Nuclear Materials," New York, NY, 1993.
B&PV Code	American Society of Mechanical Engineers, "ASME Boiler and Pressure Vessel Code," New York, NY, 1998.
NUREG-0612	U.S. Nuclear Regulatory Commission, "Control of Heavy Loads at Nuclear Power Plants," NUREG-0612, National Technical Information Service, Springfield, VA, July 1980.
NUREG/CR-3019	U.S. Nuclear Regulatory Commission, "Recommended Welding Criteria for Use in the Fabrication of Shipping Containers for Radioactive Materials," NUREG/CR-3019, U.S. Government Printing Office, Washington, D.C., March 1984.
NUREG/CR-3854	U.S. Nuclear Regulatory Commission, "Fabrication Criteria for Shipping Containers," NUREG/CR-3854, U.S. Government Printing Office, Washington, D.C., March 1985.

APPENDIX A – STANDARD REVIEW PLAN CORRELATION WITH 10 CFR PART 71 AND REGULATORY GUIDE 7.9

The following table summarizes the correlation of the SRP review procedure sections with the appropriate sections of 10 CFR Part 71 and RG 7.9.

Table A-1 Standard Review Plan Correlation with 10 CFR Part 71 and Regulatory Guide 7.9.

SRP Review Procedure Section	10 CFR Part 71 Section	RG 7.9 Section
1.5.1	None	Introduction
1.5.2	71.13, 71.31(a)(1), 71.31(a)(2), 71.31(a)(3), 71.31(b), 71.31(c), 71.33(a)(1), 71.33(a)(3), 71.35(b), 71.37, 71.38, 71.59, 71.107(c)	1.1
1.5.3	71.31(a)(1), 71.33(a)(2), 71.33(a)(4), 71.33(a)(5), 71.33(a)(6), 71.33(b), 71.43(b)	1.2
1.5.4	71.31(a)(2), 71.35(a), 71.41(a)	None
1.5.5	None	1.3
2.5.1	71.31(a)(1), 71.31(c), 71.33	2.1, 2.2
2.5.2	71.43(d)	2.3, 2.4
2.5.3	71.45	2.5
2.5.4	71.31(a)(2), 71.35(a), 71.41(a), 71.61, 71.71, 71.73	2.6, 2.7
2.5.5	71.35(a), 71.41(a), 71.43(f), 71.51(a)(1), 71.55(d)(4), 71.71	2.6
2.5.6	71.35(a), 71.41(a), 71.73	2.7
2.5.7	71.61	None
2.5.8	71.85(b)	None
2.5.9	None	2.10
3.5.1	71.31(a)(1), 71.31(c), 71.33(a)(5), 71.33(a)(6), 71.33(b)(1), 71.33(b)(3), 71.33(b)(5), 71.33(b)(7), 71.33(b)(8), 71.51(c)	3.1
3.5.2	71.31(a)(1), 71.33(a)(5)	3.2, 3.3
3.5.3	71.31(a)(2), 71.35(a), 71.41(a)	None

SRP Review Procedure Section	10 CFR Part 71 Section	RG 7.9 Section
3.5.4	71.43(g)	None
3.5.5	71.43(f), 71.51(a)(1), 71.71	3.4
3.5.6	71.73	3.5
3.5.7	None	3.6
4.5.1	71.31(a)(1), 71.31(c), 71.33(a)(4), 71.33(a)(5), 71.33(b)(1), 71.33(b)(3), 71.33(b)(5), 71.33(b)(7), 71.43(c), 71.43(d), 71.43(e)	4.1
4.5.2	71.31(a)(2), 71.35(a), 71.41(a), 71.43(f), 71.43(h), 71.51(a)(1), 71.51(c)	4.2
4.5.3	71.31(a)(2), 71.35(a), 71.41(a), 71.51(a)(2), 71.51(c)	4.3
4.5.4	None	4.5
5.5.1	71.31(a)(1), 71.31(c), 71.33(a)(5)	5.1
5.5.2	71.31(a)(1), 71.33(b)(1), 71.33(b)(2), 71.33(b)(3)	5.2
5.5.3	71.31(a), 71.31(b)	5.3
5.5.4	71.31(a)(2), 71.35(a), 71.41(a), 71.43(f), 71.47(b), 71.51(a)(1), 71.51(a)(2)	5.4
5.5.5	None	5.5
6.5.1	71.31(a)(1), 71.31(c), 71.33(a)(5), 71.35(b), 71.59(b)	6.1
6.5.2	71.31(a)(1), 71.33(b)(1), 71.33(b)(2), 71.33(b)(3), 71.83	6.2
6.5.3	71.31(a)(2), 71.35(a), 71.41(a)	6.3
6.5.4	71.35, 71.43(f), 71.51(a)(1), 71.55(b), 71.55(d), 71.55(e)	6.4
6.5.5	71.35, 71.59	6.4
6.5.6	71.35, 71.59	6.4
6.5.7	71.31(a)(2), 71.35	6.5
6.5.8	None	6.6

SRP Review Procedure Section	10 CFR Part 71 Section	RG 7.9 Section
7.5.1	71.31(c), 71.35(c), 71.43(g), 71.47(b), 71.47(c), 71.47(d), 71.87, 71.89	7.1
7.5.2	71.35(c)	7.2
7.5.3	71.87(i)	7.3
7.5.4	71.35(c)	None
7.5.5	None	7.4
8.2.4	71.31(c), 71.37(b), 71.85(a), 71.85(b), 71.85(c), 71.87(g), 71.93(b)	8.1
8.3.4	71.31(c), 71.37(b), 71.87(b), 71.87(g), 71.93(b)	8.2

APPENDIX B – TABLE OF EXTERNAL DOSE RATES FOR EXCLUSIVE-USE SHIPMENTS

The following table summarizes the information that should be provided by the applicant for the external dose rates for transportation packages for spent nuclear fuel.

Table B-1 External Dose Rates for Packages (Exclusive-Use Shipment).

Normal Conditions of Transport			
Package Surface[a]			
Radiation	**Top**	**Side**	**Bottom**
Gamma[b]			
Neutron[c]			
Total			
10 CFR 71.47(b)(1) Limit[d]	2 (200)	2 (200)	2 (200)
Package Surface[e]			
Radiation	**Top**	**Side**	**Bottom**
Gamma[b]			
Neutron[c]			
Total			
10 CFR 71.47(b)(1)(i-iii) Limit[d]	10 (1000)	10 (1000)	10 (1000)
Vehicle Outer Surface[f]			
Radiation	**Top**	**Side**	**Bottom**
Gamma[b]			
Neutron[c]			
Total			
10 CFR 71.47(b)(2) Limit[d]	2 (200)	2 (200)	2 (200)

Table B.1 (Cont.) External Dose Rates for Packages (Exclusive-Use Shipment).

Normal Conditions of Transport			
2 Meters from Vehicle Outer Surface[g]			
Radiation	**Top**	**Side**	**Bottom**
Gamma[b]			
Neutron[c]			
Total			
10 CFR 71.47(b)(3) Limit[d]	0.1 (10)	0.1 (10)	0.1 (10)
Normally Occupied Positions in Vehicle[h]			
Radiation	**Top**	**Side**	**Bottom**
Gamma[b]			
Neutron[c]			
Total			
10 CFR 71.47(b)(4) Limit[d]	0.02 (2)	0.02 (2)	0.02 (2)
Hypothetical Accident Conditions			
1 Meter from Surface of Package			
Radiation	**Top**	**Side**	**Bottom**
Gamma[b]			
Neutron[c]			
Total			
10 CFR 71.51(a)(2) Limit[d]	10 (1000)	10 (1000)	10 (1000)

[a] External surface of package.

[b] Gamma dose rate based on _____ MWd burnup, ___ % ^{235}U enrichment, and ___ years cooling time.

[c] Neutron dose rate based on _____ MWd burnup, ___ % ^{235}U enrichment, and ___ years cooling time.

[d] Dose rate in mSv/h (mrem/h).

[e] External surface of package provided that shipment is in a closed vehicle, package position within vehicle remains fixed during transport, and no loading or unloading operations occur en route.

[f] At any point on the outer surface of the vehicle, including the upper and lower surfaces; or in a non-closed vehicle, at any point on the vertical planes projected from the outer edges of the vehicle, on the upper surface of the load or enclosure (if applicable), and on the lower external surface of the vehicle.

[g] At any point 2 meters (80 inches) from the outer lateral surface of the vehicle (excluding the top and underside of the vehicle); or in the case of a non-closed vehicle, at any point 2 meters (6.6 feet) from the vertical planes projected by the outer edges of the vehicle (excluding the top and underside of the vehicle).

^h In any normally occupied space, except that this provision does not apply to private carriers if exposed personnel under their control wear radiation dosimetry devices in conformance with 10 CFR 20.1502.